# IFIP Series on Computer Graphics

*Editors*
J. L. Encarnação
G. G. Grinstein

S. Cunningham   R.J. Hubbold (Eds.)

# Interactive Learning Through Visualization

The Impact
of Computer Graphics in Education

With 94 Figures

Springer-Verlag
Berlin  Heidelberg  New York
London  Paris  Tokyo
Hong Kong  Barcelona
Budapest

Dr. Steve Cunningham
Department of Computer Science
California State University Stanislaus
Turlock, CA 95380, USA

Dr. Roger J. Hubbold
Department of Computer Science
University of Manchester
Oxford Road
Manchester M13 9PL, UK

ISBN 3-540-55105-0 Springer-Verlag Berlin Heidelberg New York
ISBN 0-387-55105-0 Springer-Verlag New York Berlin Heidelberg

This work is subject to copyright. All rights are reserved, whether the whole or part of the material is concerned, specifically the rights of translation, reprinting, reuse of illustrations, recitation, broadcasting, reproduction on microfilms or in any other way, and storage in data banks. Duplication of this publication or parts thereof is permitted only under the provisions of the German Copyright Law of September 9, 1965, in its current version, and permission for use must always be obtained from Springer-Verlag. Violations are liable for prosecution under the German Copyright Law.

© Springer-Verlag Berlin Heidelberg 1992
Printed in Germany

The use of registered names, trademarks, etc. in this publication does not imply, even in the absence of a specific statement, that such names are exempt from the relevant protective laws and regulations and therefore free for general use.

Typesetting: Camera ready by author
33/3140 - 5 4 3 2 1 0 - Printed on acid-free paper

# Contents

1 Introduction
  *Steve Cunningham and Roger Hubbold*

## Multimedia and Hypermedia

9 Electronic Books and Interactive Illustrations – Transcript of a Talk
  *Andries van Dam*

25 Opportunities for Multimedia in Education
  *Richard L. Phillips*

37 Mnemotechnics and the Challenge of Hypermedia
  *John Lansdown*

49 Cooperative Learning Using Hypermedia
  *Christoph Hornung*

65 HyperGraph – A Hypermedia System for Computer Graphics Education
  *G. Scott Owen*

79 Hyper-Simulator Based Learning Environment to Enhance Human Understanding
  *Shogo Nishida*

## Visual Thinking and Visualization

91 Visual Thinkers, Mental Models and Computer Visualization
  *Thomas G. West*

103 The Multi-Faceted Blackboard: Computer Graphics in Higher Education
  *Judith R. Brown*

115 Remarks on Mathematical Courseware
  *Charlie Gunn*

129 Visual Ways of Knowing, Thinking, and Interacting
  *Kenneth O'Connell*

137 Visualization of Concepts in Physics
  *Hermann Haertel*

145 Prospero: A System for Representing the Lazy Evaluation of Functions
  *Jonathan P. Taylor*

## Classroom Experiences

- 161 Computer Assisted Lecturing: One Implementation
  *Jacques Raymond*
- 173 Interactive Computer Graphics via Telecommunications
  *Joan Truckenbrod and Barbara Mones-Hattal*
- 189 Collaborative Computer Graphics Education
  *Donna J. Cox*
- 201 Portability of Educational Materials Using Graphics
  *Bernard Levrat*
- 211 Collaboration between Industry and Academia – Computer Graphics in Design Education
  *Adele Newton*
- 217 Computer Graphics in Computer Graphics Education
  *Ahmad H. Nasri*
- 227 Solid Modeling in Computer Graphics Education
  *Alberto Paoluzzi*

## Working Group Reports

- 243 Working Group Reports
- 244 Visual Learning (Visual Literacy)
- 250 Exploitation of Current Technology to Improve Learning
- 255 Computer Graphics as a Tool in Teaching
- 259 Long Range Views of Computer Graphics and Education
- 263 Working Group Participants

## Appendices

- 265 I – Participant List
- 269 II – International Programme Committee
- 271 Index

# Introduction

This book contains a selection of papers presented at the Computer Graphics and Education '91 Conference, held from 4th to 6th April 1991, in Begur, Spain. The conference was organised under the auspices of the International Federation for Information Processing (IFIP) Working Group 5.10 on Computer Graphics.

The goal of the organisers was to take a forward look at the impact on education of anticipated developments in graphics and related technologies, such as multimedia, in the next five years. We felt that at a time when many educational establishments are facing financial stringency and when major changes are taking place in patterns of education and training, this could be valuable for both educators and companies developing the technology: for educators, because they are often too bogged down in day-to-day problems to undertake adequate forward planning, and for companies, to see some of the problems faced by educators and to see what their future requirements might be.

## Refereed Papers

We decided first that the conference should be a working conference, in which all those attending would be expected to write position papers and to present them where appropriate to the programme. Nobody would get a free ride! An International Programme Committee was established to review papers, and a Call for Papers was distributed internationally. We received a total of 56 papers, of which 24 were selected and their authors invited to participate in the conference. All of the papers accepted were included in the proceedings distributed at the conference. A subset of the papers, including all of those presented at the conference, are included in this book. The participants, including multiple authors and organisers, are listed in Appendix I, and the International Programme Committee is listed in Appendix II.

## Invited Presentations

We invited several speakers who are well known internationally, to give keynote papers which would help us shape the programme around our intended themes. These invited papers are also included in this book.

The speakers were:

- **Andries van Dam** from Brown University, who is a well known pioneer educator and researcher in computer graphics, early hypertext, and on-line teaching systems. Andy discusses the development of truly interactive and intelligent books. Rather than providing only navigation through static material, as in many current multimedia systems, future machines will support true simulations and 3D interfaces, allowing their users to ask "what if?" questions within a multimedia context. This will require integration of diverse technologies, and the paper gives a good overview of some of the problems which will need to be solved.

- **Richard Phillips** from Los Alamos National Laboratories, who is a leading authority in multimedia systems and a former teacher at the University of Michigan. Dick describes his MediaView system, which was convincingly demonstrated at the conference by showing interactive excerpts from the book *Computer Graphics Principles and Practice* (Foley, van Dam, Feiner, and Hughes, Addison-Wesley, 1990).

- **R. John Lansdown**, Professor in the Faculty of Art and Design at Middlesex Polytechnic, who as a former architect, founder of a leading London software consultancy, and educator who uses multimedia technology in his teaching, brings a truly multi-disciplinary viewpoint. John highlights some of the difficulties which will need to be addressed before multimedia systems can be expected to achieve their full potential in education.

- **Christoph Hornung**, from the Fraunhofer Computer Graphics Group in Darmstadt, whose research centres on future on-line teaching systems for large-scale computer-based teaching and training. Christoph describes the scale of change taking place in education and training in Europe, and describes the features and structure of future cooperative hypermedia systems.

- **Bernard Levrat**, Director of the Computer Centre at the University of Geneva, who has been involved in the development and use of computer-based teaching systems for many years, but who is also at the sharp end of having to plan and budget for the equipment needed to support these activities. Bernard deals with real practical issues, such as the lengthy timescales needed for software and courseware development compared with the rapid evolution of hardware.

# Introduction

## Working Groups

The conference was scheduled to include considerable discussion time, so that people from very diverse backgrounds would have a chance to exchange views and identify common problems and approaches to solving them. In order to provide some structure for this, a number of Working Groups were established to discuss and produce reports on key areas. Their reports are included in the last section of the book. The areas covered were:

- **Visual Learning.** Some people are good at abstract thinking, others at visual thinking. If we expect to make much greater use of computer graphics in future computer-based teaching systems, what can we learn about how to make the most of visual thinking abilities? This is an especially difficult area to describe, but one which nonetheless warrants attention.

  The group decided that traditional course structures tend to reinforce either verbal literacy or visual literacy, but rarely both. Much creative work involves the use of sketching as a means to externalise ideas and make them concrete and communicable. This can be very valuable when learning, and seems to be a skill which the visually literate possess, but current computer graphics software is inadequate in its support for this. There is a need to address this area. Similarly, teaching systems require the ability to generate intelligent explanations. The group noted the serious differences in resources in different countries that used to be referred to as the North-South divide, but which now encompasses, for example, the differences between Western and Eastern European countries.

- **Exploitation of Current Technologies to Improve Learning.** Educational establishments cannot afford to throw away their existing systems every time that a new technological development arrives. The goal for this group was to examine how new teaching methods can be introduced with current systems.

  The group concluded that the problem is not only a question of investment in hardware, serious though this may be. It was noted that there is an innate conservatism amongst many educators. In part, this may be explained by the lack of reward for courseware development. In many countries, advancement of one's career depends on success in research, not teaching. The effort required to develop high-quality courseware considerably exceeds that needed to write a textbook, so there is little incentive for academics to undertake this unless attitudes change. The problem is compounded because educational software is not viewed as profitable by software suppliers. To address some of these problems a number of properly funded centres need to be established (akin to the supercomputer centres in the U.S.A.) to undertake development of suitable systems and courseware. A radical change in attitudes by higher education establishments will be needed, with changes in the way students are assessed and greater emphasis on self-learning rather than traditional lectures.

- **Computer Graphics as a Tool in Teaching.** With a wide variety of disciplines, countries, and cultures represented at the conference, it was felt that we should look for common threads between subject areas. Are diverse disciplines very different in their needs, or are there really common elements?

    It was clear within the group that the use of computer graphics in teaching is minimal. Most software that is used is developed locally by individual teachers, with very little of this finding its way to other centres. There was a consensus that many different subjects could benefit from interactive teaching software, which was seen as a major new way to open up educational opportunities and to motivate students to learn at their own pace. In subjects like engineering or chemistry it would be possible to simulate experiments that would otherwise be very costly. Software would need to be adaptable, both for the teacher and the student. Again, the issue of cost arose with regard to the expense of developing good courseware, and it was felt that the only cost-effective way to tackle this would be to set up development centres, requiring national and international funding. It should be possible to incorporate multi-lingual support so that systems can be used effectively in different countries, leading to economies of scale.

- **Long Range Views of Computer Graphics and Education.** This group was nicknamed the "revolution group." Their charge was to ignore current systems and to attempt to predict what the ideal educational workstation will look like five years hence.

    The group was confident that advances in technology will provide machines with the raw power to support advanced learning and simulation. From a hardware standpoint we can expect typical student workstations in five years' time to support multimedia, stereoscopic display, sound, and speech input and output. The major problem which remains unsolved is how to integrate all these components and to derive standards which will support interchange of interactive courseware. Among other things, systems will need to support the linking of sound and video with simulations, multimedia electronic mail, and desktop computer supported cooperative work (CSCW). There appears to be a strong concensus that a major shift will be required in graphics away from subroutine libraries towards object-oriented approaches. The potential of these can already be glimpsed with systems like MediaView.

## Demonstrations of Educational Software

We encouraged participants to bring demonstration systems with them for others to see and try. Macintosh, Silicon Graphics and NeXT workstations were available at the conference and several participants brought their systems to the conference for others to see and experiment with.

## Acknowledgments

We would like to acknowledge here the enormous contribution made by our co-organisers. José Encarnação, Head of the Computer Graphics Centre in Darmstadt, Germany, Ken Brodlie, of the University of Leeds, England, and Pere Brunet, of the Polytechnic University of Catalunya, Spain.

José is Chairman of IFIP WG 5.10, and it was at his invitation that this conference was organised. Between them, José and Ken persuaded many of the leading workstation manufacturers to sponsor our conference, making it possible for us to organise an event which academics could afford to attend!

Pere Brunet was the local organiser in Spain. Not only did he find us a superb location, but with his staff — Anna Ibanez, Emilia Bordoy, Dolors Ayala and Isabel Navazo — he managed to ensure that everything ran smoothly. This is no mean feat when it involves importing machines from abroad and getting them delivered to a rocky peninsula 100 miles from Barcelona, as well as all the usual administration of running a conference with international participants. Our thanks go to all of them for their enthusiasm and professionalism.

Without the direct financial sponsorship of several organisations, this conference would not have taken place at all. It is a pleasure to acknowledge the support and help of ACM SIGGRAPH, Digital Equipment Corporation, EUROGRAPHICS, the Fraunhofer-Arbeitsgruppe für Graphische Datenverarbeitung in Darmstadt (FhG-AGD), Hewlett-Packard, the Poltechnical University of Catalonia, Siemens, Silicon Graphics, the Spanish Government, Springer-Verlag, Sun Microsystems, and the Zentrum für Graphische Datenverarbeitung in Darmstadt (ZGDV).

*Steve Cunningham & Roger Hubbold*
September 1991

# MULTIMEDIA AND HYPERMEDIA

# Electronic Books and Interactive Illustrations — Transcript of a Talk

## Andries van Dam

Someone who thinks, speaks, and writes about the same subject for over twenty years might be seen as lacking imagination. However, I prefer to view this behavior on my part as simple Dutch persistence and dedication to a particular vision, the realization of which, I have been saying all this time, is imminent. And today, I am pleased to observe, electronic books are even more imminent, indeed, inevitable, owing to some exciting breakthroughs in both computer power and the way computers are used. The things we'd really like to be able to do are unfortunately not yet commercial reality. Therefore one of the things I want to focus on is the R & D agenda. Specifically, I want to define three remaining problems that must be solved before some of the techniques I envision can find their way into everyday computing, particularly into computing in education, which has been my special interest for well over twenty years.

Before talking about electronic books, we should review the advantages and disadvantages of paper books. We all know the good things about books: they are compact and portable, they are attractively printed, easy to read, cheap to reproduce, and, intrinsically, they are multimedia. We can annotate them, and we certainly know how to navigate, to find our way about, in books. However, searching by linear page-flipping, even with an index or table of contents, is pretty limited. And, while individual books are compact, complete libraries are not. Furthermore, in libraries there are no economies of scale: costs in larger libraries increase at least linearly with size. Maintenance is difficult and expensive: it costs more to maintain a book in a library for three or four years than to acquire it in the first place, and that cost continues while the book becomes more and more obsolete. And, perhaps most importantly, books are static, non-varying, and instantly out of date. A book, therefore, is in no way a live medium; it is a dead recording which cannot be altered to suit particular tastes or needs. A textbook, for example, is written for a specific level, such as ninth graders or college freshmen, and if you don't fit the specific profile, the book won't satisfy your needs. What is needed is a text capable of adapting itself, its contents and its presentation, to the user's needs. Electronic books will have such capabilities, and will also allow assured access, circumventing current library problems of limited copies, loss, misshelving, theft, deterioration, and mutilation. Furthermore, text searching can be expanded to include searches based on image, video, or audio sources.

My goal is to preserve the best features of paper while taking full advantage of the dynamics of the computer medium. For me, then, an electronic book would be a live database, in the full sense of that term. It would be capable of continuous updating, and would have access and version controls, etc. It would be portable, interactive, adaptable, and multimedia, capable of presenting the same material in various formats, according to user preference.

Am I therefore advocating a paperless society? I don't think that is desirable or possible. Let me read you from a 1964 article entitled "The Banishment of Paperwork," by AI pioneer Arthur Samuel: "In 1984 libraries for books will have ceased to exist in the more advanced countries, except for a few which will be preserved in museums, and most of the world's knowledge will be in machine-readable form." He also predicted automatic language translation, the translating telephone, and a number of other technologies still under development. In no way do I believe that we will or should get rid of libraries and various print technologies as they exist, but I want to widen the choices. For rapidly changing subjects, I maintain that electronic publishing is clearly preferable.

But aren't we doing a lot of electronic book publishing today? Not in the sense in which I am using the term. I am talking not about systems for producing printed material, but about something actually more like a bulletin board, an informal medium that allows people to communicate in real time. A primitive example is Videotex, which is not doing very well for a number of reasons, primarily its low resolution, low bandwidth, and low level of interactivity. You can branch through a decision tree by menu picking, and that's about all. Videotex exemplifies a vicious-cycle problem: because the medium is poor, people are not interested in writing for it, therefore you don't have good materials, and consequently there is no particular eagerness to use the medium. It is used for stock quotations, weather reports, airline reservations, and teleshopping, but it shows little promise for wider use, and certainly not for educational applications. And, although the Sears/IBM Prodigy System in the U.S. has nearly a million subscribers, it is a trivial medium, not very exciting from an educator's point of view. Students simply cannot be expected to endure presentations on Videotex systems: the screens are just not fast enough, dynamic enough, or of high enough quality.

As regards multimedia applications, various technologies can be employed: videodisc, optical disc, CD-ROMs, etc. There are also various video compression/decompression systems, but these technologies have not yet paid off and are still in minimal production use. What we will see is increasing use of them as distribution media; for example, many of us are now getting our new software releases from manufacturers on CD-ROMs and, in the financial industry, stock-market data are released on CD-ROMs. For browsing and searching technical documentation and for traditional CAI, these media are quite attractive, and I think they will be used increasingly in static, non-customizable information systems. Some technical materials, scientific journals, and possibly some textbooks will be designed specifically to be displayed and used on-line. These multimedia technologies will certainly be useful, but they do not move far enough in the direction I am advocating.

Another simple precursor to electronic books, Hypercard (and its clones), is very popular now, at least in the United States. Even though relatively primitive, it has undeniably influenced the educational field. It has been used primarily in basic high-school and grade-school courses, although it is beginning to be used for somewhat technical material such as introductory biology and chemistry. However, Hypercard applications are restricted to a card format, and in addition linking is unidirectional and allowed only between cards; you can't point to a region in another card. Navigation is also quite primitive, and Hypertalk, both a strength and weakness of the system, is not easy to use and is not a full-fledged programming language. Nonetheless, millions of people have been introduced to primitive notions of hypermedia by Hypercard and its lookalikes. Many operating systems now provide online Help using simple hypertext links from menu entries and "hot" terms.

I claim that the limitations of these media tend to lock us in and make it difficult for us to make radical progress. In addition to hardware and metaphor limitations, another type of limitation is also impeding progress: too many materials are being produced by people who have not been properly trained. While they may know about programming, they all too clearly know little about effective communication in an intrinsically graphic medium. Another limitation is the ergonomics of display screens. Most people still like to read from paper and will continue to do so until higher-resolution anti-aliased screens with good contrast and no glare are developed. Furthermore, the materials available in all these media are accessible only in review mode; they are not interactive. You have basically the ability of a VCR tape player: you can go backwards and forwards, possibly to an arbitrary frame. That is not true interactivity; one wants to interact directly with the material, not merely with a canned sequence. Another problem is the lack of standards. That none of these media interconnect seriously limits their acceptance in our field. Finally, authoring in multimedia or hypermedia is extremely time-consuming, because the authoring tools are still very primitive and also because authoring a coherent, integrated multimedia document is intrinsically much harder than authoring for individual media.

Professor Raj Reddy of Carnegie Mellon University recently gave a very interesting review of hardware development. He predicts that in the 1990s we will have "3G" machines: a giga-instruction per second, a gigabyte of memory, and a gigabyte bus. (Of course, for graphics we also will require a million antialiased, textured triangles per second or their equivalent in surface patches to produce compelling dynamic 3D displays.) Reddy also discussed the baseline "SILK" technologies he felt would be available for the user interface on such a platform: Speech, Image, Language, and Knowledge. Speech will be available for input, output, and voice annotation, and even for some recognition, so that oral communication with the system, in addition to typing or handprinting, will finally become possible. Many people, especially in Japan, are working on that idea. Reddy also foresees an ability to handle large quantities of image data (because of the high bandwidth and high compression ratios he feels will soon be available) and he predicts that the long-sought language ability, so important for language translation, will be available in the coming decade. (I am somewhat sceptical about that, except in limited domains of discourse.) Because of a built-in knowledge base, Reddy feels that it will be possible for systems to react more intelligently and to tolerate some degree of error and ambiguity in human-computer dialogue.

What will be the form of the electronic book that will exploit 3G machines and SILK-based technology? My vision is that it will be a dynamic, multimedia database that can exist in a dynabook package (a la Alan Kay) as well as in a wall-sized (flat) display. It will be based on three technologies: a hypermedia framework, a knowledge base for intelligent user interfaces and content-based dialogue between user and system, and user-controlled realtime animation at the nodes. The hypermedia issue is largely solved, except for the horrendous problems of standards and interoperability. My research group and I are concentrating on the third of the above-mentioned technologies, animation. Such animation is not merely playback of prerecorded materials, but is real-time animation generated on the fly from a manipulable model under user control. We are attempting to attain the same levels of parametric control with three-dimensional materials as are now available in a spreadsheet, whose basic formulae and data can be controlled in real time.

At this point, it should be stated that, especially in education, the distinction between authoring and delivery systems should not be too finely drawn. Readers and students should be authors too; a sense of participatory democracy should determine who has the right to put materials into the database. Authoring will, I believe, remain the most difficult problem we will face, and we must do everything in our power to make it as easy as possible for all classes of users.

I will not discuss electronic libraries here, except to point out that at the level of these technologies, the distinction between a book and a library, to a great extent, simply disappears. For example, how large is a book? To how many other books is it linked? In the limit, of course, all the information anywhere should be readily available. I won't discuss the storage and access problems associated with libraries of such enormous size, but rather will turn now to a brief review of hypermedia.

## Hypertext and Hypermedia

This field of research was defined by three visionaries: Vannevar Bush, who wrote about the Memex in a 1945 article entitled "As We May Think"; Ted Nelson, who coined the word "hypertext" and wrote in the mid-sixties about its uses in education; and Doug Engelbart, who invented the mouse and implemented the first, and still one of the most powerful, hypertext systems, called NLS or Augment.

Presidential Science Advisor Vannevar Bush, even though he predated computer technology, had very good ideas about what one wants to do with information. He coined the term "information explosion" in 1945, and predicted the development of "thinking machines" in an *Atlantic Monthly* article, later revised for *Life* magazine. In the *Life* article, Bush described voice dictation systems and other technologies that modern researchers are still developing. Bush's key concept, the Memex, was a device for reading, recording, and annotation. It functioned through the establishment of what he called "associative trails," essentially linked lists of pointers that connect materials with some logical association. He was mirroring the way he assumed our brains organize information, that is, associatively and not linearly. Because of the need to handle images, he conceptualized the Memex as a microfilm-based device. Once on film, pages would be stored in the Memex microfilm repository, along with annotations, and associative links would then be established from sources to destinations. It is this type of information linkage that is the essence of hypertext, and generalizes footnotes and "see also" cross-references.

In the mid-seventies, Ted Nelson wrote an evocative book called *Dream Machines/Computer Lib*, recently reissued by Microsoft Press in a slightly revised and updated form. It contains a lot of blue-sky thinking, some of it pretty unstructured, but also a lot of really good ideas. It shows you how far back a few creative people were thinking about interesting and provocative ways of using text, graphics, and databases, especially for education. One of Ted's many aphorisms is "if computers are the wave of the future, then surely displays are the surfboards." I have always believed that.

Hypertext/hypermedia is essentially a directed graph structure with nodes and links, and some descriptors on those links. The nodes can be any medium and of any size, that is a piece of text, a picture or a piece of a picture, a sound bite, a video sequence, whatever you like. If the node is too large to fit on the screen, appropriate scrolling controls are provided. Links can be static or dynamic; they can be based on content or on structure; they can be automatically based on search criteria, or the author can point-and-click to indicate them; they may be untyped or of different types so that you can make selected types appear and disappear to declutter the display. All links are based on sticky or persistent selection; that is, you designate some region, typically a two-dimensional region (although it could be a region in time), as either the source or the destination of a link. These selections, often called anchors, ought to be of arbitrary size, ranging from points without area to subsets of a document or entire documents. The links ought to be able to connect any type of anchor to any other type, and they can be unidirectional or bidirectional — for navigation, especially for traveling backwards, it is vital that links be bidirectional.

I will briefly outline two examples of hypermedia systems with which I am familiar. One is Brown University's Intermedia System and the other is a commercial product designed primarily for technical documentation called *Dyna*Text. Intermedia is available under Apple AUX on the Macintosh II, and it has text, graphics, tables, and timelines, as well as support for video and audio. It also has very fancy text-searching capabilities, in the sense that it can even do morphologically-based searching. So if you ask for the term "organize," it can get you a number of variants of that root such as "organization," and that turns out to be very useful. Text searching is still very important, even in a hypertext system, because jumping around using links is not the only way in which you want to get at information.

Intermedia has arbitrary-sized source and destination regions, links are bidirectional, and linking uses a cut-and-paste-style metaphor. Four actions are required: click on a selection, select "start link" from a menu, make another selection, and click on "complete link" from the menu. Now you have a permanent link between those two selections. Intermedia's most powerful feature is that the entire directed-graph structure, called a "web," is stored in a distributed database so that multiple readers can have access to it. (Only a single writer can alter the web so that there are proper access controls.) The key point is that superposition of the web does not alter original materials at all. This separation of original materials and web also makes it possible to have multiple webs at the same time, which is extremely convenient in an educational setting. For many collections of materials there are different interpretations; such pluralism of points of view is very important and can be captured in a number of ways. One is that we can have multiple webs open simultaneously, and you can see how different people connect materials in different ways.

Intermedia's major drawback is that, like every other hypertext and hypermedia system, it exists in its own private world that does not connect to anyone else's world. Thus, although hypertext technology has come a long way, it is severely limited in applicability, because it is all done within closed systems. We need open systems in order to be able to connect the world's knowledge together. More about this later.

Now let me describe Electronic Book Technology, Inc.'s *Dyna*Text system. As you know, many desktop publishing systems today can produce compound documents that contain text and graphics and other media. Some of these systems can produce files with embedded markup codes called SGML, Standard Generalized Markup Language. SGML is an ISO-defined standard. You should think of these codes as generalized type-setting codes, except they deal not merely with formatting, but rather with the structure of a document. SGML is a way of giving the equivalent of a BNF (Backus Normal Form) for a document. A document is composed of a title and a number of chapters, each with subsections. Subsections themselves can contain subsections, paragraphs, lists of items, footnotes, images, etc., and we can represent this as a kind of parse tree that looks a lot like the parse tree for a program in a programming language. So the system encodes essentially the grammar of a compound document, all the structure that a document of that particular type can contain.

*Dyna*Text reads SGML marked-up files produced by other systems, indexes the document to build a big data structure, and then formats the file for on-line reading using style sheets that define the visual appearance, that is, the semantics, of each of these syntactic SGML tags. You thus separate the syntax of a document from the semantics or appearance-oriented interpretation, and therefore the same document can have a multiplicity of appearances and you can make it look different on the screen as a function of how you like to see things. The advantage, then, of using standard SGML is that it is declarative, not procedural, in the sense that it tells you what you are describing, but not yet how it is to appear on the screen.

SGML is hierarchical, it is formal, and it is a standard. You deal with three types of files: a document-type definition (DTD) that contains the grammar, all the different components and how they can be combined, and their tags; documents that have SGML tags defined by a DTD embedded in them; and one or more style sheets for each of the DTDs. Note that by using style sheets we control not just font, size, color of text, headings, footnotes, etc., but also choice of link icon and its placement in the document, whether in the margin or inline. The document is tagged by the author during the creation phase on whatever desktop publishing system is used and is then formatted automatically for on-line reading. To facilitate navigation, *Dyna*Text produces a table of contents that can be elided or expanded by clicking and whose entries are linked to the corresponding sections. There is also a content search facility that lists the number of occurrences next to each entry in the table of contents, to give the reader a quantitative relevance estimate. Finally, clicking on an icon can launch another application. For example, in a chapter from an on-line differential geometry text, there are links to interactive illustrations provided by a Mathematica-like package controlled by Motif widgets, including a virtual sphere control for the visualization of the curves and surfaces and sliders for the parameters of each illustration. Of course, the computer had better be pretty fast to make such controls responsive!

After this very brief description of two rather different hypermedia systems, let's return to the issue of open systems. Although Ted Nelson has been extolling the mythical "Docuverse" where all our materials are connected, very much as Vannevar Bush did 45 years ago, we actually have "Docuchaos" because nothing connects to anything else. Norm Meyrowitz, former head of the Intermedia team, proposed three years ago that linking protocols be embedded in the operating system so that at the most fundamental level of the software there would be an understanding of persistent selections of source and destination regions and links between them. All applications could then use these link services. We are finally starting to see such link services with such facilities as DDE and OLE for Microsoft Windows 3.0, which allows various kinds of links. The clumsy mechanism that does this does not make these links lightweight enough for my taste, but it is at least a start and I think we are finally going to see applications such as spreadsheets, presentation graphics and word-processing programs take advantage of the DDE facility to link to each other. A similar thing is happening under OS 7.0 on the Mac with the Publisher/Subscriber mechanism. Maybe even UNIX will catch up one of these days! The problem, of course, is that all these link protocols will be incompatible. There are no standards, so even when you have linking protocols you will still, in the best of cases, have to go through a file conversion to go from one format to another; in the worst case you won't be able to at all. It will be next to impossible to link from materials in a PC environment to materials in a Mac environment, even when the machines are connected via networking, and that's criminal, in my view. Locally open systems exist, but we need to solve some really hard technical problems on wide-area links and link services and so on. That calls for much better understanding of global databases than we have today.

Portability is a more easily solved issue for electronic books — we want to carry them around, and that is starting to be possible with the latest generation of Notebook computers. No implementors, however, worry enough about size of the database. The documentation for modern weapons systems can weigh about as much as the weapon systems themselves and can occupy more volume. Also, how can we talk about databases without such elementary things as version control, which doesn't exist yet in the hypertext/hypermedia world? These systems are very primitive from an industrial-strength point of view.

The next problem is structure, or the lack of it. Most of you are very familiar by now with the fact that hypertext is just "go-to's" by another name, that if you don't watch out you get graphs that look like spaghetti code, and that navigating in that mess means getting "lost in hyperspace."

We need to design what I have called structured hypertext. How do we induce some discipline on these materials that we create? Once again, that's where we need graphic designers to help us. To minimize the lost-in-hyperspace syndrome, one can selectively elide information and show only links of a certain type. We can use various kinds of maps, for example, to tell us where we are: concept maps at very high levels of abstraction, global maps, local maps, maps that show where you've been and where you can go next, and so on. An Intermedia unit done at Brown in a geology course for exploring the moon has video tied into it as well, and you can select regions on the moon and get information on them. There is a time line, and a chronological trail is generated automatically by the system that tells you where you've been and how you got there, that you got here by opening a document, there by following a link, here by doing a content search, etc. Then, given the fact that you are at a certain node in the graph, it displays all the nearest neighbors with different link types, to tell you something about them.

So I am asking for graphic design discipline, not just to deal with the proper use of colors and fonts, but also about the best use of links. It's a new medium, with an evolving design discipline, so we cannot get too impatient as long as we recognize that it needs an awful lot of work. After all, it took centuries for book design to evolve, and decades for all the other media we know about. Dick Philip's MediaView system is a tasteful prototype of a good hypermedia system that takes excellent advantage of both the aesthetics of the NeXT platform and its object-oriented, multimedia capability. It would be a fine system on which to develop the "elements of style" for a new hypermedia design discipline. And English and Art History professor George Landow at Brown has been developing design guidelines for what he calls the "rhetoric" of hypertext.

## Artificial Intelligence

The second basic technology of the electronic book is AI. What do I mean by AI? You would like your books to know something about their own contents, to be able to interact with them, ask them questions, have them understand something about your goals, to answer questions in your context, in terms of what you know and what you don't know. You'd like to be able to let the books lay themselves out on the screen using some design rules. You'd like to be able to have autonomous agents find things for you, agents that you need not dispatch explicitly, but have been programmed in advance to know something about you, the kinds of things you want to do and your tastes in exposition. While you are working at your workstation or when you are away from it, they compose your personalized electronic newspaper, your hypermedia document. So when you walk in in the morning the system has already found the things that happened last night in the news, in the world of politics, commerce, science, entertainment, whatever you are interested in, and formatted it for you. These agents have gone around and examined sources, done information retrieval of a variety of intelligent types, and together with layout agents composed your document. This is a very powerful example and sounds still a little like science fiction, but I believe it will be with us well within this decade. NewSpeak at MIT, a decade ago, was able to do much of this scenario! Bob Kahn and Vince Cerf, who for years were driving forces at DARPA, have now founded the Corporation for National Research Initiatives, a nonprofit research outfit that is studying the use of ultra-highspeed, gigabyte-per-second networks, and autonomous programs they call Knowbots that survey these networks looking at various sources of information, from earth satellites to news feeds, gathering information and presenting it to you in the form you would like it to have. Another good example of the use of AI technology is automating authoring. Steve Feiner and Kathy McKeown at Columbia University are making progress in the automatic generation of graphics and textual annotation for stylized documents such as maintenance and repair manuals that use procedural, step-by-step instruction.

## User-Controlled Animation/Interactive Illustrations

My research group and I have been concentrating for the last five years or so on user-controlled, model-driven, real-time animation, the third basic technology for electronic books. I don't mean simply playing back prerecorded sequences of batch animation. Such prestored animation is very good for a large number of subjects in which you do not have to control the content of the presentation. But in an educational setting, when you want to explain the structure of a molecule or of a steam engine or a particular mechanical linkage or some abstraction in mathematics, you do not want to be restricted to a canned set of video clips because you will never be able to prestore all the different points of view with all the different parameters to satisfy any particular user. It will be too much like a textbook where the author has had to make a selection from many possibilities, and you choose one among these prior selections. The student should be able to look at the object with the level of detail and the parameters that he or she wants, completely customized. This means that you need to be able to derive these illustrations in real time from stored models, analogous to the spreadsheet experimentation I mentioned earlier. The user controls the parameters, which in turn control the structure and behavior of the model, and also how you view things with the software's synthetic camera — hence my term "interactive illustration."

I think such model-driven animation presents us with a "grand challenge." Right now animation is done by trained professionals who really have to know a lot to be able to do an animation of the type we are used to in canned video form. Ordinary people should be able to construct visualizations in a relatively short amount of time, so that instead of a project that you must spend months on, it is something that you can do in a day or two. Even professors should be able to do it! There is no such thing as being able to do a simple animation in an afternoon today. There is no equivalent to MacDraw or MacPaint for animation. You can do a little key-frame interpolation, or a film loop from prestored images, but that is not what I am talking about. Authoring an animation means creating a time-varying model and all its behaviors. There are no commercial systems designed to do this today, only systems for producing batch animation; but we, among others, are working on research prototypes. For the special and simple case of 2D real-time algorithm animation, Brown has developed reasonable tools (BALSA, TANGO) that give you a vocabulary of simple shapes and movements. Little animations of rectangles moving smoothly to show, say, sorting algorithms can be put together by a computer science student perhaps in a day, although that is still far too much work. However, as far as I am concerned, 2D animation is a solved problem for simple shapes. So let's talk about the really hard stuff, animations that are far more complicated animations.

3D animations are incomparably harder. The objects specified are usually much more complex and the changes you want over time are much harder to specify. Adding a time dimension is implicit in most of what we do, and it is very tricky. We are not very knowledgeable, except perhaps in the world of music, on how to treat time as an explicit dimension. And, of course, we are dealing with a four-dimensional entity (three-space plus time) on a two-dimensional screen, and that presents a very difficult problem of indirect manipulation.

So we can talk about building a model of a microworld of interest, simulating the behavior of that model, and then animating whatever results from rendering it. But there is no single magic-bullet technique that does the job; we need a collection of techniques, including keyframe animation, scripted animation, gestural animation, kinematics and inverse kinematics, dynamics and inverse dynamics, dynamic constraint satisfaction, procedural modeling, etc. You'll notice many of these are techniques from Newtonian physics. Indeed, there is a movement in computer graphics which I characterize as the "back-to-physics" movement, in which Newtonian mechanics has become very important. If we want to talk about a number of objects behaving in a three-dimensional space, interacting with each other, colliding, rebounding and so on, then we are back to basic physics. How do we simulate that physics fast, and how do we make it look good enough for display purposes, although it might not be exact physics? And the most difficult research problem is not just to do these techniques fast, but to integrate them. That's one of the areas that we've chosen to do our research in at Brown.

Real-time animation based on simulation is an extremely resource-intensive problem and I claim that we need desktop supercomputers, not merely good workstations. Raj Reddy's 3G machine will be just about right (at least for now!). I believe you cannot do anything very interesting in this arena on PCs, Macintoshes, Sun Sparcstations, Personal Irises, etc. Today's PCs and workstations, even those with hardware support for 3D graphics, are all just interesting enough to show us what could be done, but not good enough to do it yet. You need many hundreds of mips and, more importantly, many hundreds of megaflops to do scientific computation, especially simulation and animation in real time. Also, you need far more display power than is currently available on relatively cheap workstations.

Well, the good news is that this power is coming. If you look at the newest generation of HP machines, for example, you get from fifty to seventy mips, as much as twenty honest Linpak megaflops, not peak megaflops, for prices ranging from $10,000 to $30,000. I think that is remarkable, and I think that gives us, again, a giant step forward in what you can put on your desktop. Therefore, when we talk about 1995, with two more generations of hardware/software improvement, we can certainly look for a factor of four, if not more, in price/performance improvement. Which means that today's $20,0000 workstation might only cost $5,000 and have more power by a good bit than even that fifty mips that you can get today for approximately $10,000. So I don't think it is crazy at all to talk about a

"Crayola" on your desk by the end of the decade, and really mean something that is Cray 2-style performance in a "pizza box," the form factor that people like to deal with today.

The problem, as usual, won't be hardware but software. I think an awful lot of fundamental work is going to be required before a collection of disparate real-time/animation techniques come together in a sort of uniform whole. What we have today is batch animation done by professionals; there is almost no equivalent to "clip art" for animation. Swivel-3D on the Mac is a start. What we need is a large and growing collection of three-dimensional models with time-varying behaviors that the user can specify or at least control. We need national projects, large projects, like the human genome project or fusion, something as significant and as profound, with as much financial and human support, to build a family of models for our study of objects and phenomena. These should be upward-compatible families, in some sense, so that, for example, when fifth graders study atoms and molecules they get a very simplified model, and then when they go to high school we add a little more complexity, a little fine structure, if you like. But we all know that this is still an approximation to what you learn in first-year university chemistry, which in turn is an approximation to what you learn when you know something about quantum phenomena, and that in turn is replaced by even more robust and realistic models for the working scientist. Simpler models should lead to more complicated models, and the student should be able to see how the more complex model generalizes or adds features that were not in the simpler model. Thus students could progress, on their own or with supervision, from a high-school model to a university model and perhaps beyond, without being stuck in tracks predetermined for age and experience levels, restricting their imagination.

Most of the time people talk about the output, i.e., the renderings or the simulation, of a model, but I think input is an important and neglected area. We need just as much work in the input arena as we do in the output side and on the physics side because, ultimately, we are still talking about interactive illustrations, and that means we need good tools with which to interact. The mouse, which is basically a two-dimensional device, and the keyboard and the stylus are not the appropriate devices for much of what we want to do in 3D.

The University of North Carolina specializes in serious uses of virtual reality, for molecular modeling or building walk-throughs, in particular. They use a force-feedback joystick with six degrees of freedom, and that, combined with real-time stereo, gives you a very nice virtual reality presentation. You can, for example, do molecule docking maneuvers, taking some group and trying to make it fit into various receptor sites, with visual as well as tactile feedback. I think virtual reality, even though it's distressingly hyped for its science-fiction aspects, has very interesting applications and we should definitely be watching that field. Right now the stereo goggles are still too crude and the amount of movement and degree of realism of the images are still too small, but the state of the art is improving nicely and tripling or quadrupling processor speed will make virtual reality simulation a lot more pleasant and a lot easier in the next few years.

To summarize, I think we are looking at a tremendously powerful new medium, because we're really looking at a database, with very many nice user interface properties. Documents will be live, they will be interactive, they will have some built-in intelligence. If we get the right tools it will mean that we can produce and disseminate information not only far more quickly but also far more cheaply, and provide far more universal and democratic access to information than we do now. Maybe we can start removing some of the boundaries between the information "haves" and "have-nots." Even in the highly developed countries you can see huge gaps between those who have immediate access to information and those who do not, and I believe that if we have good economies of scale and make future Dynabooks as inexpensive to own as television sets, which they will be, even with supercomputer power, then we will have a fantastic tool for giving more universal access to interesting information.

Many hard problems have to be overcome, many of them not even technical. What do intellectual ownership and property rights mean in this new age? Who owns what? Who is an author? Who is a reader? How is the author compensated for the reader's use of the material, and so on? These are very difficult issues because we're dealing with an exceedingly plastic medium that is a collection of multiple media and we know almost nothing about how to work with such complex, composite media. Also, how easy it is to be superficial and to dazzle someone with beautiful graphics and sound! People walk away and say "I loved it," but when you test, you find out they learned essentially nothing, absorbed nothing. The users were amused, and that's not all the medium should be about. We want to grab their attention, but we also must be sure that we're going to transmit some real information, some knowledge and understanding, and how do we do that? How do we make sure that the glitz, in fact, does not take away from the substance of what is to be learned?

Back to technology for a moment. While the right technology is just around the corner, we still must solve some very hard technical problems before the full vision of electronic books can be realized. Among these are time-based modeling for real-time, user-controlled animation, and knowledge modeling for language understanding, agents, content retrieval, automatic authoring, and intelligent dialogue. In the United States, fortunately, there is now tremendous commercial pressure to create simple electronic books because the government, in particular through the CALS (Computer-Aided Acquisition and Logistics Support System) initiative, has decreed that all information for DOD weapon systems must be available in electronic book form. And there are other industries, such as the aircraft and telecommunications industries, that also see the need for doing that. I believe we will finally reach the first truly interesting and societally influential level of development in this decade when we have color Dynabooks with supercomputer performance, SILK interface technologies, solutions to the problem of linking protocols over wide-area networks, and multi-paradigm, real-time animation for interactive illustrations.

# Opportunities for Multimedia in Education

## Richard L. Phillips

If approached correctly, multimedia technology can provide enormous benefits to education, especially higher education. Before elaborating on that statement, we should have a working definition of the term "multimedia." Literally, of course, it means communication using more than one medium. For our purposes, the term implies a computer-based device which, in addition to a textual display, typically has a graphics capability, voice and music output and a live video display. An application programmed for such a system can orchestrate the invocation of each of these media at the appropriate time. Very effective presentations can be produced which include animated sequences, sound and text. Programs written for the Macintosh under HyperCard exemplify such applications.

### Competing Approaches and Pitfalls

The opening sentence suggests that there might be an incorrect way to embrace multimedia capabilities for education (indeed, we argue, for any serious use). We are at a crossroads now where the path we take can determine how successful multimedia will be. We have on the one hand the computer industry adding sound and video capabilities to workstations and on the other the consumer electronics and telephone companies adding computing capability to their products. In a recent article [1], Negroponte comments "Looking back on this period, historians from the next millennium will wonder why these two business and engineering groups did not talk to each other." And they are *not* talking to each other, or even listening when the other speaks.

What we have is a Philips Electronics describing multimedia in terms of the Compact Disc Interactive (CD-I) format and associated television peripherals that will make use of it. We have a standards group populated by telephone company representatives who have adopted the px64 standard for communication of low bandwidth digital video. These are both typical of several *least common denominator* approaches which, if they are widely adopted, will undermine the potential of multimedia for years to come.

## Computers Must Lead the Way

It is critical that a multimedia system be first and foremost a computer, not an appliance or an augmented telephone. Approaches being taken by the computer industry offer the greatest power, flexibility and promise for serious use. If multimedia is to make inroads in education, it is important that we recognize faulty approaches and reject them.

## Shortcomings of Current Software

Even the computer industry is not living up to its potential. While there are some computer companies who are seriously pursuing the development of multimedia hardware and software, resulting systems, for the most part, are only *multi-mediocre*. Educators are urged to eschew these systems and hold out for something better. Let's name names — Apple and IBM; HyperCard and ActionMedia. Both offerings, as well as those from less well-known companies, suffer from a variety of shortcomings. Among them are:

- *Expert orientation* – to produce anything useful or powerful the author must be a programmer.
- *Author orientation* – the systems are more useful to the author than the end user, a subversion of the goal of multimedia.
- *Inflexibility* – an author's agenda and strategy for interaction is rigidly imposed on end users, who may wish to explore information differently.
- *Unfortunate metaphors* – note cards and page turning are inconvenient strategies for learning.
- *Lack of user empowerment* – documents cannot be dynamically created or modified; the cut, copy, and paste paradigms are unavailable in these systems.

If a multimedia system is to succeed in education, all of the above objections must be overcome. At this point there is no sign that the major vendors recognize this or are willing to change their approach. Later we will discuss a prototype system developed by the author that embodies the correct approach to multimedia. This system is called MediaView.

## Shortcomings of Current Hardware

Today's hardware is simply inadequate to the task of serving education's needs in multimedia. We dismiss immediately the Philips/CD-I system as an ill-conceived toy. But what do other companies offer us? CD-ROM and its variations, analog video from laser discs and digital video interactive, DVI.

### *Problems of Size and Flexibility*

What's wrong with this? First, a CD-ROM based system is immediately limited by a bounded dataset, which is also incapable of being modified. True, 750 MB of data on a small disk *seems* unbounded but when one includes 24-bit images, sound and digital video sequences, that space is used very quickly. If all the data is textual, significantly large corpora can be stored there and could serve as a starting point for a system that allows dynamic construction of multimedia "documents."

### *Analog Video*

Systems that accommodate only analog video require not only a peripheral disc or tape player, but often a separate monitor for viewing the video. A greater disadvantage, though, is that the video is not really part of a multimedia "document." It exists outside the system and it cannot be edited or participate in copy/paste operations. In short, analog video is inconvenient and inflexible. What we want is digital video.

### *Digital Video, but Wrong*

DVI is digital video but it comes with its own set of problems. The current offering from Intel (and IBM) is an asymmetric scheme, meaning that the decompression process prior to viewing is different from the compression done on frames captured from analog video. In fact, users are expected to send their video sequences to a "compression factory," much like sending a film away to be processed. So, it is impossible for the user to *create* digital video dynamically, an important part of the empowerment issue raised earlier. In addition to these shortcomings, the format of DVI is ill suited for transmission over digital networks. This, we will see, is important if we are to take full advantage of the promise of multimedia.

## What is Needed?

There are emerging technologies, both in hardware and software, that offer the opportunity to make multimedia systems powerful, flexible and useful in education. Until these technologies are incorporated into systems, however, there is little available that is worth the expenditure of any educational resources, financial or human.

### Hardware Features

Standards are emerging that deal correctly with digital video. JPEG [2] is here, and soon there will be MPEG, a form of JPEG that accounts for temporal coherence. There are already chip vendors with products that support JPEG and we will soon see them appearing in systems. In fact, NeXT Computer, Inc. has already announced NeXTdimension [3], a board system that supports several formats of analog video and offers symmetric JPEG compression and decompression for full motion video.

The CD-ROM should be replaced with read/write optical discs. Such discs are on the market now but there exists no standard format or size. Committees are at work now to specify these standards, but it will probably be a year or more before the medium is widely available. Once they are, users can create and modify their own datasets. But the size of these discs, probably 100 MB, will still be insufficient for most serious applications. What is needed are networked multimedia servers, which can appear to the user to be effectively infinite in size. More on this later.

### Software Features

One can probably infer a list of desirable features of multimedia systems from the shortcomings enumerated earlier for existing software. To be specific, they are as follows.

- The system must cater to the non-expert, not just the end user but the author as well. Both groups must be able to easily educe the maximum potential of the system.

- The end user of a system ought not to be constrained by the agenda of the author. That is, full and random exploration of a "document" should be possible, regardless of a syllabus the author might have had in mind.

- It should be possible to modify existing "documents" dynamically and to create new ones from them easily. This is facilitated by applying familiar

what-you-see-is-what-you-get (WYSIWYG) word processor interaction paradigms to all multimedia components. Just as words are cut, copied and pasted, so too should video fragments, images, sound, and the like.
- Because multimedia systems are designed for communication, it should be possible to share "documents" electronically. This means that they can be shared with a co-worker on a local network or mailed electronically to a colleague at a distant site. Animations and any other substructure of the "document" should persist through the mailing process and be awakened at the destination by the recipient.
- The most important attribute of a multimedia system is *explorability*. Once a concept has been introduced, the end user should be free to investigate, say, to simulate a chemical process or to render a synthetic scene. In effect a live laboratory should be available to the user, otherwise a multimedia system offers nothing but a novel way of reading a textbook. The importance of this feature underscores the earlier contention that multimedia systems should be based on a platform that is primarily a computer. Moreover, it emphasizes the futility of the television peripheral approach being pursued by Philips, for example.

A system embodying these features and principles has been developed by the author. It is called MediaView and is described below.

## MEDIAVIEW, a Step in the Right Direction

MediaView is a multimedia digital publication system that was designed to be flexible and free from policy. It was also designed to take maximum advantage of the media-rich hardware and software capabilities of the NeXT [4] computer, especially the features of the NeXTdimension [3] sub-system. Rather than emphasize the word that is almost always paired with multimedia, *presentation*, MediaView emphasizes *communication*. The system does not tacitly impose the "author's" agenda on the reader, but structure and sequencing can be overlaid if appropriate. What has resulted is an extremely general system, one that is free of artificial structure and inconvenient metaphors, such as note cards, page turning, and the like.

### Features of MediaView

MediaView is easy to use and understand. It is based on the WYSIWYG word processor metaphor, something familiar to most computer users. In addition to text, that metaphor is extended to include all multimedia components. Like text, these components are subject to the select/cut/copy/paste paradigm, making them as simple to manipulate as words. As a result, powerful and complex MediaView documents can be constructed by non-specialists.

In addition to the expected multimedia components like graphics, audio and video, MediaView supports several non-traditional components. These include full color images; object-based animations; image-based animations; mathematics; and custom, dynamically loadable components. In providing such a range of capabilities, MediaView fully exploits the platform integration and media richness of NeXT, NeXTstep, and NeXTdimension. Indeed, MediaView was first developed for this environment because it is a precursor of systems of the future.

Finally, being designed for maximum communicability, MediaView allows multimedia documents to be electronically mailed to remote sites. In short, MediaView is a communication tool that offers new and dramatically different ways of interacting with others. A full description of MediaView and some of its applications appears in [5].

## *Applications to Education*

While MediaView enjoys broad applicability it has some features that are especially useful in education. These can be demonstrated by a recent project aimed at presenting selected chapters of a well-known computer graphics textbook [6] in a multimedia format. For this project three novel multimedia techniques have been developed. They all exploit the explorative, live laboratory capability proposed above. They are best described by showing images derived from interacting with the pertinent chapters.

## *Exploring Algorithms*

Readers who are familiar with [6] are aware that many of the computer graphics algorithms presented appear in the text in the form of pseudo-code. In the MediaView version of the text this code is "live," and can be explored. Figure 1 is a composite image, showing the code fragment for the DDA scan conversion algorithm for lines and a sketching window, where the user can try out different lines and watch the algorithm perform. The advantage over the conventional textbook is that the reader can explore concepts further than space and the medium permitted the authors to do. Rather than be content with one or two selected static images, readers can try as many parameters as they choose.

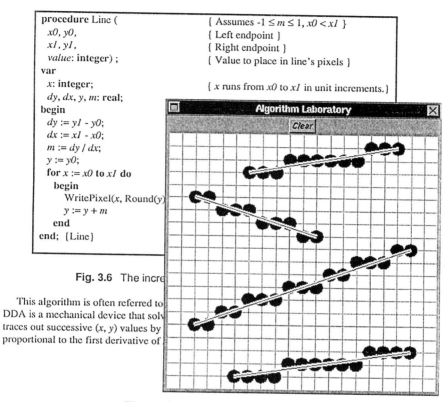

```
procedure Line (                    { Assumes -1 ≤ m ≤ 1, x0 < x1 }
    x0, y0,                         { Left endpoint }
    x1, y1,                         { Right endpoint }
    value: integer) ;               { Value to place in line's pixels }
var
    x: integer;                     { x runs from x0 to x1 in unit increments.}
    dy, dx, y, m: real;
begin
    dy := y1 - y0;
    dx := x1 - x0;
    m := dy / dx;
    y := y0;
    for x := x0 to x1 do
        begin
            WritePixel(x, Round(y));
            y := y + m
        end
end; {Line}
```

Fig. 3.6 The incre

This algorithm is often referred to
DDA is a mechanical device that solv
traces out successive (x, y) values by
proportional to the first derivative of

**Figure 1: Exploring an algorithm**

## *Exploring Mathematics*

Since Mathematica [7] is bundled with NeXT computers, it is readily available. The Mach operating system on the NeXT platform allows applications to easily communicate with one another through programmable ports, which is how MediaView talks to the Mathematica application. Alternately, standard UNIX pipes can be used to communicate directly with the Mathematica server, bypassing its front end. Both approaches are used by MediaView.

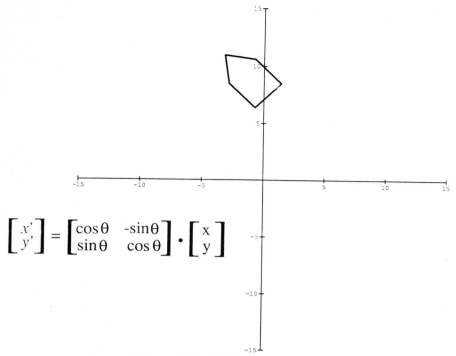

**Figure 2: Exploring an equation**

Equations that appear in a MediaView document have a backing format that is compatible with the Mathematica language. By clicking on an equation, its semantics are conveyed to Mathematica, wherein the user can perform a variety of symbolic and numerical operations. A user can directly invoke a "one-shot" operation, such as evaluating or plotting an expression, or can incrementally build up a Mathematica notebook, which can be manipulated through the standard Mathematica front end. For example, Figure 2 shows a matrix expression from Chapter 5 of [6] which describes a geometric transformation on a graphical object. Each element of the matrix can be changed to evaluate the effect of the transformation. Clicking the overall equation evaluates the expression and plots the accompanying figure.

## *Exploring a Three Dimensional Dataset*

Chapter 6 of [6] deals with viewing in three dimensions, always a difficult topic for students in computer graphics. For the first time they are exposed

# Opportunities for Multimedia in Education 33

This same result can be obtained in many other ways. For instance, with the VRP at (8, 6, 54), as in Fig. 6.30, the center of projection, given by the FRP, becomes (0, 0, 30). The window must also be changed, because its definition id based on the VRC system, the origin of which is the VRP. The appropriate window extends from -9 to 9 in $u$ and from -7 to 11 in $v$.

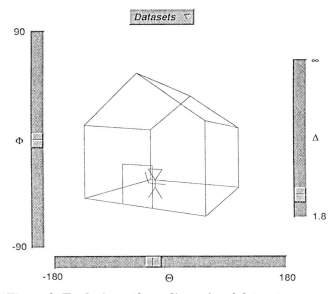

**Figure 3: Exploring a three-dimensional dataset**

to projection planes, vanishing points, view volumes, etc. Lots of verbiage and many (static) illustrations later a student still needs practice to really understand the concepts. In the MediaView version of Chapter 6 a viewing laboratory appears *in situ*. The student can select a dataset from several provided and interactively explore the effects of many three dimensional viewing parameters. Figure 3 shows a bit of text from the chapter, the layout of the laboratory and its controls.

The dataset is a slight variation on one used in the book. The F and Q sliders control rotation about the $x$ and $y$ axes, respectively, while the D slider changes the perspective effect by modifying viewing distance. For simple datasets, changes effected by the sliders are instantaneous. For more complicated datasets, say with 500 or more lines, it takes a fraction of a second to show requested changes. Incidentally, the image of any orientation for any dataset can be copied from the viewer and pasted into another document. Thus, a portfolio of customized static images can be easily produced.

## What's in the Future?

MediaView is seen as a precursor of the kind of multimedia system that will be useful in education. It is flexible, powerful and offers both teacher and student advantages that are not now available. The capabilities of such a system will be further augmented by several emerging technologies. Because of advances in digital video [8], we will enjoy desktop video conferencing over conventional baseband networks. Imagine how convenient it will be for a professor to hold office hours this way.

Also, advances in workstation window systems will give us the ability to remotely share workspaces. This means that student and teacher can graphically or textually edit the same multimedia "document" over the network. The changes made by each will be immediately visible, just as if they were both using the same conventional chalkboard. This capability, coupled with video conferencing, is in many ways superior to a personal visit.

Work is progressing on multimedia databases and servers. Companies like Oracle and Sybase are exploring ways to extend their current offerings to accommodate multimedia components. One company, Xanadu Operating Company [9], has already developed a powerful multimedia, hyperlinked database system. It is not yet available but is scheduled to begin beta testing soon.

Finally, to provide effective server support to multimedia systems we need to be able to move large amounts of data quickly. Fortunately, recent advances in network technology will soon make this feasible. There are currently several testbed projects to evaluate wide area gigabit networks, some spanning thousands of miles. Once joined to these networks, universities will have access to limitless amounts of data.

## *Acknowledgements*

It is a pleasure to acknowledge the support and encouragement of all the authors of *Computer Graphics: Principles and Practice,* and the help of Keith Wollman of Addison-Wesley Publishing Company in making original data from the book available.

## References

[1] Negroponte, N., "Vanishing Point," *NeXTWORLD* 1(1), January/February, 1991, 98–100.
[2] Wallace, G. K., "Overview of the JPEG still picture compression algorithm," *Communications of the Association for Computing Machinery* 34(4), April 1991, 30–44.
[3] NeXT Computer, Inc., *NeXTdimension*, Publication No. N6030, September 1990, Redwood City, CA.
[4] Clapp, D., The *NeXT Bible: Hardware and Software Systems for the NeXT Computer*, Brady, New York, 1990.
[5] Phillips, R. L., "MediaView: A general multimedia digital publication system," *Communications of the Association for Computing Machinery* 34(7), July 1991, 74–83.
[6] Foley, J. D., A. van Dam, S. K. Feiner, and J. F. Hughes, *Computer Graphics: Principles and Practice*, 2nd edition, Addison-Wesley, Reading, MA, 1990.
[7] Wolfram, S., *Mathematica: A System for Doing Mathematics by Computer*, Addison-Wesley, Reading, MA, 1988.
[8] Liebhold, M. and E. M. Hoffert, "Towards an open architecture for digital video," *Communications of the Association for Computing Machinery* 34(4), April 1991, 113–116.
[9] Tribble, D., et al., "FeBe Protocol 88.1x," Xanadu Operating Company, October, 1988, Palo Alto, CA.

# Mnemotechnics and the Challenge of Hypermedia

## John Lansdown

Underlying the current enthusiasm for teaching by means of multimedia and hypertext — subjects that we can term collectively, hypermedia (Figure 1) — there seem to be some implicit assumptions which need to be examined. Typical of these are

- assumptions about perception; for example, if a picture or video extract is good on its own, it must be even better with music, voiceover, and sound effects or in company with other pictures or video extracts;
- assumptions about scale; for example, if one picture is worth a thousand words, fifty thousand pictures must be worth fifty million words; and
- assumptions about cognition; for example, because the brain works by association, we should present our information associatively.

This last assumption is, of course, axiomatic to hypermedia. Vannevar Bush, the undisputed father of the subject and whose brilliantly perceptive 1945 article, "As We May Think," [2] bears constant re-reading, put it persuasively thus:

> The human mind...operates by association. With one item in its grasp, it snaps instantly to the next that is suggested by association of thoughts, in accordance with some intricate web of trails carried by the cells of the brain. ... [T]rails that are not frequently followed are prone to fade, items are not fully permanent, memory is transitory. Yet the speed of action, the intricacy of trails, the details of mental pictures, is awe-inspiring above all else in nature. ... Selection by association, rather than by indexing, may yet be mechanised. One cannot hope thus to equal the speed and flexibility with which the mind follows an associative trail, but it should be possible to beat the mind decisively in regard to the permanence and clarity of the items resurrected from storage. (Bush 1945)

These and the many other assumptions we make are not necessarily untrue; however, because some of them are at the very heart of hypermedia, they should be more closely examined than they seem to have been to date. I am not taking the viewpoint of a critic hostile to the subject. On the contrary, I am an enthusiast for hypermedia. But I believe that we will build a very shaky edifice indeed if we don't look more closely at the foundations on which it's built.

It is also worth looking at some of the lessons that might be learned from more conventional media such as film and TV. Often the aims and objectives of film and TV makers are different from those of hypermedia designers, especially as far as education is concerned. Nonetheless, there is much of relevance that can be learned from the practice and, above all, the theory of film-making. Currently, hypermedia is not much more than a set of useful but often experimental techniques separated from the mainstream of educational thinking. We are still far from having a theory of hypermedia. Without such a theory to sustain and underpin developments, the subject of hypermedia will remain nothing more than a bywater for technological enthusiasts. It deserves to be more than that.

## *Problems of Perception*

The brain is an organism which uses massively parallel methods to do its "processing." This sometimes leads us to believe that parallelism is a feature of all cerebral activity; however, the way in which we attend to, and hence respond to, external information is not wholly parallel. Thus although we are able to walk and chew gum at the same time, only in fairly limited circumstances can we usefully take in information simultaneously from multiple sources.

But, as our hypermedia systems become more powerful and more versatile, we are already beginning to see them used in ways which ignore this simple fact. Taking their cue from certain types of TV programme, some hypermedia designers are beginning to "multiplex" information — they simultaneously present the viewer with often only tenuously related images, sound effects, music, voiceover, and still and moving text. The visual and aural effect of such presentations on TV is often exciting. Apparently — for some TV teenage audiences — it is necessary to present material in this way to persuade them to look at all. But this form of multiplexing is much better at creating mood rather than imparting knowledge and should only be sparingly applied in an educational setting.

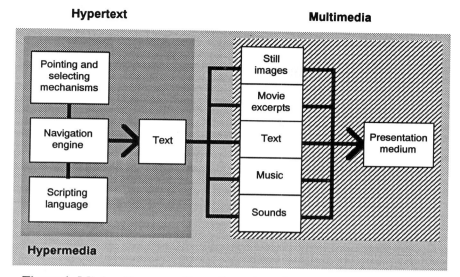

Figure 1: **Minimal hypermedia as the union of hypertext and multimedia**

That appropriate music and sound effects can enhance visual presentations is well-known to film audiences. Even so-called silent movies were virtually always shown with musical accompaniment. And this was not provided just in the form of solo piano improvisations. In the larger urban cinemas, it was quite usual to have a full orchestra playing specially composed music supplied by the film's makers. Later — but still before talking pictures came on the scene in the late 20s — musical accompaniment was provided from recorded discs played on phonographs which were synchronised to the pictures. These discs, too, were supplied by the producers. However, we have all experienced both annoyance at inappropriate or intrusive accompanying music and amusement at infelicitous sound effects. We are only too aware of the way these frequently interfere with our understanding of whatever meaning the images are trying to convey.

But sound properly related to image has another effect which many of us must have experienced but find hard to explain. If, in the night, we look across to neighbours' houses, we sometimes see their TV sets working. We are too far away to recognise the images and, of course, neither can we hear their accompanying sounds. What we see is a series of patches of coloured light moving from one state to another with enormous rapidity. The images do not look as though they could form a part of a coherent film at all, and we would find great difficulty in distinguishing what we can see in the distant

TVs from some abstract display of colours. Yet when we switch on our own TV set to hear the sounds, not only does the speed of intercutting between the distant images seem to slow down significantly, but we also seem to be able to make sense of the images themselves.

We can deduce from all this and from many studies on attention — some of which are well summarised in Wilding [15] — that presenting information on multiple channels is valuable, but only if the methods of doing it are properly thought out. We have to attend in order to perceive, and multi-channel presentation can either help focus this attention or diffuse it.

However, it has been aptly said by Moray [13] that "we have come to see attention not merely as a single process concerned with enhancing the clarity of perception. Rather, it is a complex of skills. These include selecting one from several messages, selecting one from several interpretations of information. ... But in our interactions with the rich dynamics of the world in which we live and to which we adapt, attention also models that world, and provides us with strategies and high-level control of our tactics of information sampling, optimising our information-processing in the face of our limited processing capacities." This fundamental role of attention in our understanding cannot be overlooked, and we must take it into account if we wish to present information coherently and well.

## *Problems of Scale*

One of the most impressive characteristics of the technology of hypermedia is its prodigious capacity for storage and delivery of information. By means of this technology we can potentially control and access almost unlimited levels of educational resource. In this paper we will ignore the critical problems of ownership of information in its various manifestations. Questions of copyright, intellectual property rights, and fair dealing will loom even larger in this area than they do at present, not only as we capture and translate information into new forms, but also as we bring together elements from various sources in order to transform and enhance meanings.

The sheer volume of the possible resource brings with it potential for

- the fairly obvious and, I trust, easily resisted temptation to overwhelm our students and ourselves with huge quantities of information; and
- the more difficult problem of becoming properly familiar with details of the resource so we can locate and deliver desired elements of it at will.

Although exacerbated by the new technology, this latter problem is not, of course, unique to hypermedia. All of us already experience difficulties in locating particular half-remembered images or items of text in libraries and slide and video collections, especially if they have large-scale provisions (my own educational collection, for example, runs to a few thousand slides and hundreds of animation excerpts, whilst the Middlesex Polytechnic Art and Design Faculty Library has over 350,000 slides and videos). Obviously the difficulties of location are eased by good cataloguing methods and knowledgeable librarians. But the ready availability of large image collections on videodisc and CD-ROM make the possibilities of personal cataloguing very problematical.

Take, for instance, the Smithsonian Museum double-videodisc collection of aircraft pictures. This collection contains over 100,000 images and is the sort of resource that many of us might have in our personal hypermedia libraries. To browse through this collection looking at each picture for a scant five seconds would, on the assumption of five hours browsing a day, take nearly twenty-eight days! To catalogue these images, with each one taking as little as five minutes to enter into a database, would take well over a working year.

When one's interest is not simply in named or nameable objects (such as particular aircraft), but in such things as the circumstances involved in the taking of the pictures or in the incidental social information that can gleaned from the picture backgrounds, the problems of conventional cataloguing are almost overwhelming. The London Transport Museum's set of 120,000 historical glass negatives of vehicles is a prime example of a collection that has this sort of incidental social interest. Ostensibly these images concern London trams, buses and trains over the past 150 years, but because most of the photographs show the vehicles in their contemporaneous surroundings they also present a unique social record of their times. According to one's focus of interests these images are a resource for the history of industrial design, architecture, advertising, fashion, or many other things. Unfortunately this collection is not yet available on videodisc and has already taken the Museum three years to catalogue on the basis of verbal descriptions of the pictures' foregrounds and backgrounds, a database which is a resource of enormous educational value.

When the hypermedia resource also contains video extracts, sounds, voice, music, and text in addition to still photographs and synthesised images, the problems in remembering its scope and nature so that its full potential can be realised are immense. Indeed the problems lead me to suggest that the greatest barriers to full use of hypermedia will not be those of technology but of human mechanisms for remembering. Side by side with developing hypermedia resources, therefore, we also need to develop a new discipline, one which we can call *mnemotechnics*.

## Mnemotechnics: Ways of Reminding and Remembering

In order to do any work properly, all of us need to have abilities to remember large amounts of information which, elsewhere (Lansdown [7]), I have categorised into three basic types: global, domain, and task-related.

Because some professionals handle task-related information, they are faced with problems especially relevant to those outlined above. Television and video editors in particular have to be able rapidly to recall appropriate sequences of images, sounds, and music when putting together edited highlights of news or sports transmissions. Unfortunately, almost nothing seems to be known about the mental processes these editors use when carrying out this task, and we are left with an impression of its being a magical art. Much could be learned from research into the way in which these editors work. But until this research is done we have to examine other ways in which we might improve our abilities to remember the contents and location of information sources.

One promising approach might lie in more formally reviving a mnemonic technique which is very ancient indeed. This is the so-called method of loci which seems to have been first described by Cicero in 55 B.C. and has been in use in various forms ever since (see Yates [16] for a history of the subject up to mediaeval times, and Luria [9] for a spectacular modern example of its use). The method comprises memorising items by placing mental pictures of them, the imagines, in predefined imagined locations, the loci. To recall entails mentally going in order through these locations and recognising the association between the images and the things to be recalled. Very often the locations were set in a theatre, giving rise to the magically-named Theatre of Memory (Yates [16]).

A surprising experiment is reported in Godden and Baddeley [5] which seems to add another dimension to the method of loci. In this experiment, diving club members were asked to learn lists of words either on the shore or at a depth of six metres under water. When asked to recall the lists, they had a substantial decline (forty percent) in memory if they learned words in one situation and tried to recall them in another. Thus, if the list was learned underwater, say, it was best recalled in the same situation and not on land, and vice versa. This seems to suggest that the method of loci is not just a set of formalised tricks for remembering items by situating them in imagined physical locations, but that it is based on a more fundamental phenomenon in which memory depends heavily on the learning context. It would be interesting, for example, to know if video editors experience a decline in their ability to recall content and location of video sequences when they are outside the milieu of the editing studio.

The method of loci has, of course, been used in the human–computer interface metaphor of the "desktop" which we use by placing items in folders on the screen-simulated desk. We find particular computer documents and applications partly by remembering where these are placed on the desktop. Negroponte [14] used the idea on a larger scale in his spatial data management system (see also Bolt [1], Donelson [3], as well as Herot et al, [6]). Malone [11] looked at the ways people arrange papers on their real desks — in files and piles — and found that the arrangements used were both for locating items and for reminding people to do things.

The need, then, to be familiar in detail with the vast volumes of information that can arise from hypermedia, even that held at a personal level, is paramount but not at all easy. The methods of achieving this familiarity will be partly technological and partly cerebral. As much as anything though, they will need to be societal. It will be necessary for us to evolve more cooperative and congenial ways of sharing information. We will also need new methods of classifying and categorising. Among other things, this will come about by a greater realisation that information is not neutral but depends on point of view, and that some categorisations are helpful for searching and some are not. For instance, Margulis and Guerrero [12] proposed new classification of life-forms based on DNA sequences. As far as I can tell, just three categories of living things are now being suggested: two categories of bacteria and one of everything else. Hardly useful categories for most information retrieval purposes, however accurately they might reflect the nature of life forms.

## *Problems of Cognition*

But one of the most problematical aspect of hypermedia for teaching purposes is that arising from the concept of associativity. As the quotation from Vannevar Bush at the start of this paper stresses, there is little doubt that that associativity plays a strong — perhaps even primary — role in linking our personal mental schemas into a coherent whole. Philosophers from Aristotle onwards seem to concur in this. What is not so certain, though, is whether the best strategy for imparting information and knowledge is by means of associative methods, particularly when these give rise to non-serial presentations.

There is direct evidence that some forms of associative methods are not very useful as teaching tools, for example, the 1960s teaching machines and "branching books" based on them. Not only did these devices sometimes leave students floundering in muddy backwaters remote from the mainstream of the subject in hand, they also prevented progress at varying speeds (just about every question had to be answered to allow movement from one subject to another), and they had difficulty in emphasising what in a subject was important and what was secondary.

Even more telling arguments against associative methods in teaching, however, are those which rely on indirect evidence: for example, the success over the last five hundred years of the conventional book form, and the development over many thousands of years of the narrative story style. Most books present their information serially and in an order determined by their authors. True, books can be skimmed and sampled. But, if all the information the book can impart is to be found, it must be read from beginning to end. The narrative story style — "Once upon a time such and such happened ... and then this and then that ... and they all lived happily ever after" — is embodied in most literatures whether for teaching or for entertainment. A non-serial presentation, unrelated to time, tends to be reserved for novels and similar art forms. But even non-serial novels (with notable exceptions, such as the eighteenth century *The Life and Opinions of Tristram Shandy, Gentleman* by Sterne) are a fairly modern manifestation.

It is hard, too, for students to lose their way in a book. This is not the case in works of hypertext. There the major problem is losing one's way; indeed, this problem is so fundamental that current papers on hypertext often mention it only in passing (for example, Dufresne et al [4]). The problem of navigation does not apply to the same extent if it is the lecturer who is controlling the progress through the hypermedia presentation and not the student. In this case it is assumed that the lecturer has a real or mental map of the subject space. If the students are to do the navigation, they should be provided with maps, signposts, and trail records which allow them to locate where they are, where they want to be, and how to get there. Much more research, then, is needed into the role of hypertext-like associativity in teaching and learning before we can be certain that it is the way to go.

But these comparisons with the book are not meant to suggest that the book is an appropriate paradigm for hypermedia. Books have a special place in the collective psyche. Their convenience, portability, seriality, permanence, and — very often — sheer beauty, are unique qualities as media for transferring information and knowledge. Perhaps a more appropriate paradigm is the illustrated lecture or seminar, wherein information is imparted in a more interactive, real-time fashion with progress and content determined by the response lecturers receive from their audiences.

## The Simulation Engine

I may have given the impression that hypermedia is concerned only with the retrieval, manipulation, and presentation of stored or pre-existing material. This would be wrong. Of course, for the time being such material forms the major feature of any hypermedia system, as Figure 1 suggests, but this is less likely to be the case as computers improve in their capabilities to deliver such things as real-time graphics and animation. In the case of straightforward musical examples where the timbral quality of the performance is not relevant, for instance, there is already little point in storing pre-recorded excerpts when synthesisers controlled via a MIDI interface can present live music. Thus a real-time configuration more on the lines of Figure 2 is to be favoured.

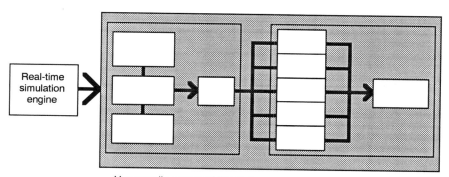

Hypermedia system of Figure 1

**Figure 2: Preferred hypermedia configuration**

It is these growing real-time capabilities of computing that present us with the greatest challenges in hypermedia. The capabilities lead us to envisage systems which can tailor their presentations to the progress needs of individual students. They will deliver stored information when needed but, by means of real-time simulation, will create new information in the form of images, music, text and voice when appropriate. The possibilities for teaching that they open up are literally endless. They will allow us properly to fulfil our roles of information providers in new and exciting ways because our role is to provide students with enriching, edifying, and enlightening information (Lansdown [8]). By providing them with information in the new and integrated forms that hypermedia makes possible, we have some hope of achieving this goal.

## Acknowledgements

My thanks to Professor Alan Mackay of Birkbeck College, University of London and Visiting Professor to Middlesex Polytechnic who brought my attention to the Theatre of Memory (see Mackay [9]). It was he and my colleague, Stephen Boyd Davis, who helped me clarify my thoughts about hypermedia and who started me thinking about some of the problems outlined here.

## References

[1] Bolt, R. A. (1979) *Spatial Data-Management*. DARPA Report, MIT, Architecture Machine Group, Cambridge MA.
[2] Bush, V. (1945) "As we may think," *Atlantic Monthly*, July, reprinted in Computer Bulletin, March 1988, pp 35–40.
[3] Donelson, W. C. (1979) "Spatial management of information," *Computer Graphics* (12) 3, pp 203–209.
[4] Dufresne, A., I. Tremblay, and S. Turcotte. (1990) "Exploratory behaviours and the design of computer instruction manuals in hypertext," ACM SIGCHI *Bulletin*, (22) 1, pp 40–41.
[5] Godden, D. R. and A. D. Baddeley. (1975) "Context-dependent memory in two natural environments: on land and underwater," *British Journal of Psychology*, (66), pp 325–331.
[6] Herot, C. F., R. Carling, Friedell, and D. Kramlich. (1980) :A prototype spatial data management system," *Computer Graphics*, (14) 3, pp 63–70.
[7] Lansdown, J. (1983) *Knowledge and Information in the Design Office of the Future*. Working Paper, System Simulation Ltd, London.
[8] Lansdown, J. (1990) "Education or training: Intellectual development or preparation for commerce?" *Proceedings PIRA/RSA Design Conference*, London.
[9] Luria, A. R. (1975) *The Mind of a Mnemonist*. Penguin, Harmondsworth
[10] Mackay, A. L. (1989) "In the mind's eye," In Lansdown, J and R. A. Earnshaw, *Computers in Art, Design and Animation*, Springer, New York, 1989, pp 3–11.
[11] Malone, T. W. (1983) "How do people organise their desks?" *ACM Transactions on Office Information Systems*, (1) 1, pp 99–112.
[12] Margulis, L. and R. Guerrero. (1991) "Kingdoms in turmoil," *New Scientist*, (129) 1761, 23 March, pp 46–50.
[13] Moray, N. (1987) "Attention," In Gregory, R. L., *The Oxford Companion to the Mind*, Oxford University Press, Oxford, pp 59–61.

[14] Negroponte, N. (1979) "Books without pages," *Proceedings International Conference on Communications* IV, (1), pp 1–8.
[15] Wilding, J. M. (1982) *Perception: From Sense to Object*. Hutchinson University Library, London.
[16] Yates, F. (1966) *The Art of Memory*. Routledge & Kegan Paul, London.

# Cooperative Learning Using Hypermedia

## Christoph Hornung

In Europe, twenty-two per cent of the population is in some form of full time education, and more than ten million people undergo some form of training or retraining every year. This amounts to eighty-five million people in education, training, or retraining each year, or one third of the European Community population [9].

The need for training and retraining is a large and growing problem. Technological change is accelerating, creating a need for more skilled workforces and resulting in substantial costs for training and retraining of employees. For the individual, this means that education will become a lifelong process rather than one concentrated in childhood and school. This, in turn, requires a different kind of organization, where individuals can work at different geographical locations: in the office or at home, during work or on vacation.

From a technical point of view, learning is a highly interactive process, requiring both immediate access to learning material and interactive communication between teacher and learner. Any system which aims to provide an environment for this, and to encompass aspects of human-human communication, must support at least voice and sound, and full motion video — in other words, it must be a multimedia system.

## State of the Art in Multimedia and Communications Technology

### Multimedia Documents

One way to describe the processing of information is to view it as the processing of an underlying *document*. We can then distinguish three main topics: the media defining the content of the document, the structures defining the internal organization of the document, and the modes of access to the document.

## Content of a Document

Although there are promising developments in multimedia, work in this area is only at the beginning. The integration of graphics or images into text is possible, but integrated editing environments are not yet available. The integration of video in a window and the support of sound is mostly technology-driven. Impressive demonstrations can be shown, but there are not yet sufficiently convincing applications of the new technology. Moreover, we are still far away from an integrated multimedia standard.

## Organization of a Document

Multimedia documents need a flexible and dynamic structure. Such a structure can be provided by links between objects in the document. A link establishes a connection between two parts of a document, but rather than being merely a reference, a link can be executable. It may be time-variant or change its value. The execution of a procedure may be associated with it, or it may be intelligent in the sense that it can *learn* during the lifetime of the document, and so on. The principal difference from a traditional document is the change from a static document, consisting of data primitives, towards an intelligent document consisting of active objects interacting with each other.

## Access to a Document

The focus of much research has been on tools to support individual work. While multiple users may have access to the resources of a computing network, they tend to work in a concurrent fashion. This means that all the users compete for the resources which the system distributes fairly among them; however, they are still effectively working separately within the system. The possibilities to interact with each other are rather spare and frequently — as with different forms of mail — not interactive.

The need for cooperative access to documents is recognized and there is ongoing research in this area. One important form is so-called joined editing, where several users have read access to an open document at the same time. Communication channels between different users are established, and they can interact by using multiple cursors, each one under the control of an individual user. This shared document thus provides a conferencing mechanism, which can be used to support joint decision making. However, in this scenario it is usual for one of the users to take the master's role. The master is the only one allowed to make modifications to the document, thereby avoiding conflicting or inconsistent accesses. Future work will concentrate on allowing multiple users to update a document at the same time, thereby establishing a new *conference computing* paradigm.

*Communication Technology*

In the past, communication technology and information technology were quite separate fields. Common examples from the telecommunications area are the telephone, telefax, and videophone. Today, however, there is convergence. There is a trend towards digital telecommunication networks and the integration of services. Integrated Services on a Digital Network (ISDN) provides an example, although its bandwidth (64 Kbits/s) is insufficient to support time-critical applications. However, broadband networks, such as DQDB or ISDN-B, will provide a bandwidth in the range of at least hundreds of Mbits/s making it possible to transmit full-motion video using image compression techniques. On the other hand, applications requiring lower bandwidth will be able to share capacity, using configurable protocols such as ATM, thereby reducing costs. Research in this area is being conducted under the European program, Research and Development in Advanced Communications Technologies in Europe (RACE).

Notwithstanding this progress, it should be noted that more work will be needed to improve the bandwidth delivered by the network to applications. Most demonstrations show the performance of the network by the transfer of full motion video. But this is only one aspect of teleconferencing. The real integration of information technology and communication technology is still an unsolved problem and is of strategic importance. The European Community is now sponsoring research in the so-called area of telematics, which is a synonym for the integration of telecommunication and informatics.

# Trends in Education and Training and System Requirements

In the field of education and training several shifts are taking place. Particularly important is the shift from teaching to learning. Exploratory learning, supported by intelligent tutoring systems and available in distributed learning environments, will become standard. Another clear shift is towards flexible learning during work, blurring the distinction between learning and working. In this section we look at different learning models

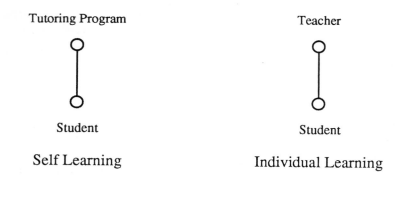
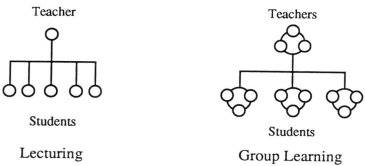

**Figure 1: Learning models**

which are emerging and identify system requirements which arise from these (Figure 1).

## *Self-learning*

Self-learning is defined by the interaction between a learner and a teaching program. The program serves as an electronic teacher and provides tutoring, monitoring, and help. Programs for self-learning are becoming more popular, especially in combination with multimedia presentation and interaction. We can find them, for example, in museums. These systems allow interaction and in this way support exploratory learning.

The teaching program itself should be adaptable and intelligent to provide optimal learning success. The user interface must be configurable and a variety of alternatives encompassing different media for the presentation must be provided. This is necessary to implement learner-focussed instead of the teacher-focussed learning. This means that the teacher is responsible for providing a variety of possible representations, but the learner, assisted by the teacher or an intelligent teaching program, can choose the one best suited for him.

*Individual Learning*

This method can be characterized by the interaction of a single learner with a single teacher. In the first instance, this form of learning provides the same functionality as self-learning. But the addition of interaction between the learner and the teacher depends on the conference computing paradigm.

Both the learner and the teacher have access to the same learning material. They can discuss several topics; the teacher can supervise the learner and give hints. To support individual teaching, visualization tools for cooperative access to learning material must be provided. A supervisor mode must be provided to allow the teacher to follow the learner's actions. Cursors must be provided for each participant to support the discussion about the material. Communication channels are needed to support the dialogue between them.

*Lecturing*

Lecturing is defined by the relation between a single teacher and multiple students. If no interaction is supported, this approach can be described as *tele-lecturing* — it is equivalent to giving a lecture by closed-circuit television. However, from a pedagogical point of view, this is not an optimal approach. For learning, interactive communication between teacher and learner is important. But if the teacher and the learners are in different places and cannot communicate naturally, as is the case in an electronic classroom, additional tools for the identification of learners and for communication between the teacher and learners must be provided.

In this situation the teacher is in a master role, while the learners follow the course and have only restricted possibilities for interaction, for example, by the use of *hot keys* to attract attention. In contrast, the teacher has the facility to navigate through the learning material, to reorganize it interactively, and to select exercises and additional examples. It may be sufficient to have a true communication connection (with sound and image) between only one selected learner and the teacher. This sound and image can then be distributed to all the other learners. In this sense, lecturing implements the same functionality as individual learning with the additional features of multiple hot key connections and distributed conference computing.

*Group Learning*

Group learning can be described as multiple connections between groups of teachers and groups of learners. This model covers both individual learning and lecturing and cooperative work. Group learning is by far the most comprehensive approach to computer-based training. Learner groups may work interactively on a common problem. This requires the provision of high-performance communication channels and decision support tools. It is essential that the participants are able to make personal annotations to a public document and to discuss them as possible alternatives in a cooperative way. Another important feature of this type of learning is the access it provides to a group of expert teachers from different areas, for a single learner, or a group of learners.

## Activities in Computer-Based Training

In the field of computer-based training, three major areas of activity can be identified: design and production, learning, training, and teaching, and administration and management.

*Design and Production*

This covers the design, development, and production of course material. Here the focus is on the definition of the course content and on its internal structure together with sets of exercises. This part should be separated as far as possible from the presentation of the course material. But the design and production of electronic courses has to be arranged with the involvement of people responsible for course credits to guarantee their acceptance. This is an essential point when considering the design of courses for industrial use.

*Learning and Teaching*

This area is primarily concerned with the question of how to learn. It covers the presentation of, and interaction with, course material as well as communication between learners and teachers. Of course, work in this field will be greatly influenced by pedagogical experts. This requires the development of a platform for learning and teaching, that provides uniform learning support tools for tutoring, monitoring, and help; uniform access tools for browsing and navigation; configurable, flexible, and individual learning environments to achieve optimal learning success; the support of distance learning; and support for Computer-Supported Cooperative Work (CSCW).

*Assessment and Management*

Here the goal is to provide a flexible environment for evaluation, examination, assessment, and supervision. The focus is on the development of uniform curricula and examinations. In this field, we see another important difference compared with today's systems: no longer will we have only tools for individual work, but rather tools to teach people, to increase their skill, and to assess their training. The successful management of computer-based training requires a common evaluation and management platform, providing uniform evaluation and assessment tools.

In Europe, with its heterogeneous structure and different languages and cultures, the development of uniform curricula is a major problem which will require close cooperation with representatives of governments.

## System Architecture for a Virtual Campus

The development of a successful computer-based training system is a complex task. The system is influenced by different technological requirements, management requirements, and methods of use and access. Different user groups must be supported: courseware developers and producers, learners and teachers, and administrators and managers. Each user group will need to work with its own application system, but they will cooperate in an open environment comprising different connected learning centers. We will refer to this as a V*irtual Campus.*

The Virtual Campus represents a uniform system architecture consisting of the following components: a *common training platform* serving as a network interface, a *training information system* serving as an application interface, *application environments* serving as toolsets for application scenarios, and specific *kernels* implementing specific applications (Figure 2).

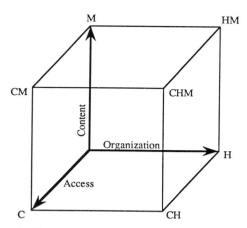

**Figure 2: System architecture for the virtual campus**

The common training platform implements a virtual computing and communication environment based on physical platforms connected via heterogeneous networks. Here the main problem is to guarantee interoperability across networks and systems. The usage of existing standards as well as the derivation of new standards for networking systems, user interfaces, and information exchange is crucial.

On the different nodes forming the common training platform, training information systems will be established. These will serve as the application platform for the virtual campus and provide a uniform look and feel and access across the network. On this level, generic tools are implemented which can be used by all user groups. These tools provide, for example, services for cooperative access, navigation through hyperstructured documents, and processing of multimedia documents.

On top of the generic tools, application-specific toolsets will be implemented geared to specific application systems. For example, a generic decision-finding tool may be customised as a cooperative authoring-support tool for courseware development, as a learning support tool for exploratory learning, or as a management-support tool for information dissemination.

On top of the hierarchy, we find small kernels implementing specific application systems for each of the applications mentioned above. These kernels are based on the application-specific toolsets using well-defined interfaces. They define a functional behaviour, while the implementation details are hidden on lower levels using an object-oriented paradigm. This allows both rapid prototyping as well as an easy exchange of applications.

# Cooperative Hypermedia Systems

The architecture sketched in the previous section is designed to support cooperative access to multimedia documents with hyperstructures. We will call such a system a Cooperative Hypermedia System (CHM). To implement cooperative learning based on hypermedia effectively, a clear understanding of the internal structure of CHM systems is necessary. As a first step in that direction, a proposal for a reference model for CHM systems will be described in the following section, characterising the different aspects of these systems [15].

## Components of a CHM System

The components of a CHM system describe the organization, content, and modes of access to a document. It is important to note that these components are considered to be independent of each other.

### Hyperstructure Organization

Documents may be organized in different ways. There are, for example, linearly structured documents consisting of segments or blocks. Tree or graph structures allow much more complex forms of organization and are often used in computer graphics and advanced document processing.

Links may be established to connect parts of different documents. Their basic use is to establish references. Hyperlinks, or intelligent links, are an extension of this concept, but a hyperlink is a method rather than a reference. This allows actions to be associated with links. Such an action may be a video sequence presented when the link is activated. A link itself is able to "learn." For example, if such a link is activated several times, the system may recognize that the user is concentrating on it and behave accordingly. This is especially useful in intelligent tutoring and monitoring systems.

In a highly structured document, a proper separation between structure and content is essential. Structure has a higher level of abstraction. A document may be structured hierarchically as a graph consisting of different structure elements, while the content of the document is defined by the leaves of this graph.

In complex documents it is easy for a learner to get "lost in hyperspace." It is therefore important to have structure visualization tools for the user to see where he is, to navigate through, or browse, the document.

One such tool would show the structural organization of the document independent of its content. This type of graph visualization tool requires the calculation of a proper layout for presentation of this information. An unresolved issue is whether a three-dimensional representation helps in the visualization of such complex structures. Another tool would support the definition of appropriate icons to visualize the structure and/or content at a higher level of abstraction.

Browsing may also be used to do exercises at a certain level of detail. If the learner successfully solves these exercises, he may switch to another topic: if he fails, he will be trained in this area with additional, more detailed material.

Another important issue is structure editing. Especially in an authoring environment, it must be possible to exchange whole chapters of a course without modifying their content. For the display of the structure graph, this requires the effective calculation of interactive modifications requiring that a new layout can be calculated incrementally.

### MULTIMEDIA CONTENT

A CHM system encompasses different media like text, graphics, video, and sound. The integration of these different media then leads to multimedia. Single media systems for text, graphics, and images are on the market and well-understood. Current multimedia systems allow the editing of documents encompassing multiple media, but further research is needed into the full integration of the different media. In particular, problems of time-variant media and appropriate user interfaces have to be considered.

The content is the most concrete aspect of a document. It is defined by the leaves of the structure graph describing the organization of the document. In a learning environment, multiple media must be provided, for both the underlying documents as well as for communication between teachers and learners. In this area, computer graphics will be used to provide tools for the integrated presentation of text, graphics, and images. Special research is needed in the area of time-variant multimedia objects.

### COOPERATIVE ACCESS

The third important aspect of a CHM system is the type of access that is allowed to a document. With the conference computing paradigm, it is essential that multiple users can interact and update a document simultaneously.

The synchronization of concurrent user requests is no longer only a system issue but becomes a part of the application, as in real conferencing. Therefore, conflicting cases may either be solved implicitly or may be recognized by the system and passed back to the users, who must solve the conflict by negotiation. Suppose that in an authoring environment one of the authors wishes to delete a section while another one wants to include something. This is clearly a semantic conflict: is the part worth modifying, or is it worthless and should be deleted? This decision cannot be made by the system but must be negotiated between the users. The system, therefore, just recognizes this situation and establishes a communication connection between the two users.

The contribution of computer graphics to the cooperation aspect is in providing appropriate tools for visualization and interaction. For each user, it must be clear which parts of the document he has access to and which ones are in use by others. This may be realized by the use of different colors [16]. In this way, many conflicting accesses can be avoided by the users themselves. Moreover, tools for the visualization of objects under negotiation should be provided. The user currently owning an object should be aware of the fact that another user wants to have access to it. The user attempting to gain access should be visually informed that his request is being processed. After the object is freed by the first user, the access right is granted to the second user. If more than two users want to access an object, access is granted in the order of the access rights.

## THE CHM CUBE

With respect to these components, a system can be described as a 3-tuple, each tuple consisting of an unordered set of features in that particular component. Hereby, a system is described as a subset of the CHM cube (Figure 3). This allows a classification of systems as well as a detailed comparison, if the features are precisely described.

## *Classification of Tools*

Having identified the components of a CHM system, the next question is, "what should a complete and well-defined CHM system consist of?" The answer is to provide tools for all the different types of hardware devices. Following this approach, we can identify tools for processing, storage, presentation, interaction, and transfer for each of the components media, organization, and interaction of a document (Figure 4).

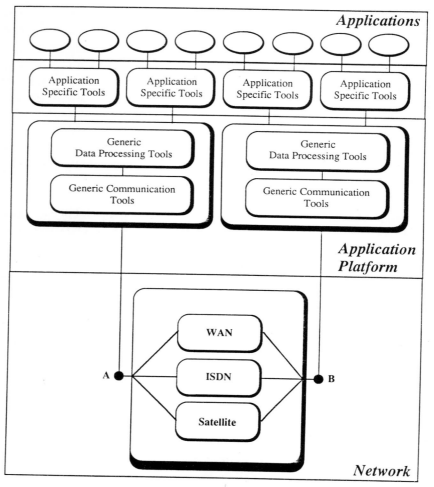

**Figure 3: The CHM cube: components of a CHM system**

## *Implementation Aspects*

What is a responsible strategy to implement a CHM system? We propose a hierarchy based on drivers, generic tools, and specific tools for each of the aspects of processing, storage, presentation, interaction and transfer (Figure 5) Special attention has to be paid to the definition of clear interfaces between the different levels and the different tools. Tools themselves can use other tools on different levels.

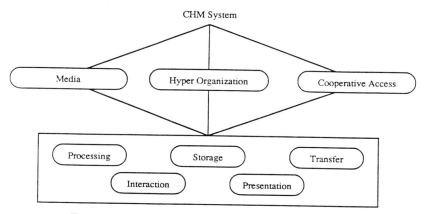

**Figure 4: Classification of tools of a CHM system**

Two more remarks are important. First, all services can be distributed and, therefore, must be designed to allow and support distribution. This means that all services must have access to the transfer services. Secondly, cooperative access to resources is essential. This makes an extension of the traditional client-server model necessary. The support of multiple clients cooperating with the resources of a single server will be called the *cooperative client-server model*.

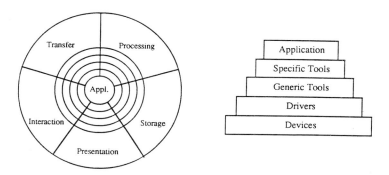

**Figure 5: Hierarchy of services in a CHM system**

## Summary

There is a strong need for improvements in Advanced Learning Technology (ALT). Skill is the power of our post-industrial society, and learning is changing from traditional education towards a lifelong, individual process. Integration of information technology and communication technologies will be the challenge of the '90s.

The Virtual Campus is a key concept for the implementation of ALT, based on a network of centers of excellence. It implements a common training platform serving as an integrated environment for courseware developers and producers, learners and teachers, and administrators and managers.

Computer-based training can be implemented as a so-called Cooperative HyperMedia system. A CHM system consists of distributed components providing cooperative access to common resources. It supports the processing of multimedia documents with hyperstructures. A modular, flexible architecture is essential to support incremental research in this new field with rapidly changing requirements.

The role of computer graphics in education will be to provide tools for presentation and interaction, including sound and video. Browsers will be required to navigate through highly structured documents. Tools for structure visualization and structure editing will be important with increasingly complex documents.

We are now at a time of fast technological change. The availability of group work applications will change our use of computers fundamentally. Computer-based education and training are among the most promising application areas profiting from the new conference computing paradigm.

## *Bibliography*

[1] Akscyn, R., D. McCracken, and E. Yoder, "KMS: A Distributed Hypermedia System for Managing Knowledge in Organizations," *Communications of the ACM*, July 1988.

[2] Barret, E., *Text, Context and Hypertext: Writing with and for the Computer*. MIT Press, Cambridge, MA, 1988.

[3] Blum, B., "Documentation for Maintenance: A Hypertext Design," Proc. Conf. on Software Maintenance, 1988.

[4] Brown, P. J., "Hypertext: the way forward," Proc. Int. Conf. on Electronic Publishing, Document Manipulation, and Typography, Nizza, April 1988.

[5] Collier, G. H., "Thoth-II: Hypertext with explicit semantics," Proc. Hypertext'87, Chapel Hill, 1987.

[6] Conklin, J., "Hypertext: An Introduction and Survey," *IEEE Computer*, 20(9), Sep.1987.

[7] Conklin, J. and M. Begeman, "gIBIS: A Hypertext Tool for Explaratory Policy Discussion," *ACM Transactions on Office Information Systems*, 6(4), Oct. 1988.
[8] Delisle, N. and M. Schwartz, *Neptune: a Hypertext System for CAD Applications*. Tektronix Laboratories, Oregon.
[9] DELTA-Workplan '91, Commission of the European Communities, Directorate General XIII, Directorate F.
[10] Durfee, E. H., V. Lesser, and D. Corkill, "Trends in Cooperative Distributed Problem Solving," *IEEE Transations on Knowledge and Data Engineering*, March 1989.
[11] Foss, C. L., "Effective Browsing in Hypertext Systems," *Proc. RIAO'88*, MIT Press, Cambridge, Mar 1988.
[12] Garg, P. K., "Abstraction Mechanisms in Hypertext," *Communications of the ACM*, 31(7), July 1988.
[13] Gibbs, S. J., *LIZA: An Extensible Groupeware Toolkit*, MCC, Software Technology Program.
[14] Hofmann, M. and H. Langendörfer, "User Support by Typed Links and Local Contexts in a Hypertext System," Proc. of the workshop "Integrierte, intelligente, Informationssysteme", Sep. 1990.
[15] Hornung, C. and A. Santos, "Proposal for a Reference Model for Cooperative Hypermedia Systems," Eurographics Multimedia Workshop, 1991.
[16] Hornung, C. and A. Santos, "CoMEDiA: A Cooperative Hypermedia Editing Architecture," Eurographics Multimedia Workshop, 1991.
[17] Draft, Scope and Purpose for New API, ISO/IEC JTC1/SC24/WG1N139
[18] Jones, H., "Developing and Distributing Hypertext Tools: Legal Inputs and Parameters," *Hypertext '87*, The University of North Carolina, Chapel Hill.
[19] Nielsen, J., "The Art of Navigating through Hypertext," *Communications of the ACM*, 33(3), March 1990.
[20] Phillips, R., "Interactive SIGGRAPH Proceedings: A New Form of Publication," *Computer Graphics*, 24(1), Jan. 1990.
[21] Phillips, R. L., "Multimedia Publications: Status, Opportunities and Problems," Proc. of the Eurographics '90 Conference.
[22] Rada, R., "Writing and Reading Hypertext," *Journal of the American Society of Information Scientists*, Mar. 1989.
[23] Rada, R. and B. Keith, "Collaborative Writing of Text and Hypertext," *Hypermedia* 1(2), 1989
[24] Ricart, G. and A. Agrawala, "An Optimal Algorithm for Mutual Exclusion in Computer Networks," *Communications of the ACM*, Jan. 1981
[25] Santos, A., "State of the Art Report on Multimedia," FAGD-90i034, Dez. 1990.
[26] Sculley, J. and I. Hutzel, "Playing Author," *Computer Graphics World*, Feb. 1990.
[27] Streitz, N. A., J. Hannemann, and M. Thöring, "From Ideas and Arguments to Hyperdocuments: Travelling through Activity Spaces," Proc. Hypertext'89, November 1989.

# HyperGraph — A Hypermedia System for Computer Graphics Education

## G. Scott Owen

The basic method of teaching has remained essentially unchanged since the nineteenth century. It consists of an instructor delivering a lecture to a group of semi-attentive students who return to their rooms to try to understand the lecture. In some fields the teachers have introduced elements of modern technology into the classroom, such as television, films, and computer demonstrations.

In this paper we suggest a drastic change in the method of delivery of instructional materials to one that fully uses modern technology, namely the development and use of a hypermedia instructional system. A hypermedia system is the combination of hypertext and multimedia computer systems. These systems allow users to read textual materials, listen to audio descriptions, view static and dynamic graphics images, stored and interactive video, and to create their own graphics images. The users can navigate through the material jumping from one area to another and are not constrained to a linear path.

Hypermedia systems [4] have enormous potential for training and education, in both scientific and non-scientific areas, but it is as yet unrealized. The development of large and complex hypermedia systems for instruction is not an easy task since there are many difficult issues to be resolved. Some of these issues are general instructional issues, some are applicable to all types of computer-aided instruction (CAI), and others are specific to hypermedia systems. For example, a general instructional issue is what subject material to cover and how to cover it, a general CAI issue is the interaction between the human and the computer, and a hypermedia specific issue is how to best navigate through the system and how to ensure that the user doesn't become lost.

We are developing HyperGraph, a hypermedia system to teach computer graphics. HyperGraph can be used both for lecture presentation and for self-paced individual student learning. Though a system to teach computer graphics is the focus of this paper, much of the discussion is also pertinent to other subject areas, especially in science or engineering technology. When fully implemented this system can be used as a prototype for similar development in other academic areas. We are developing the system using the instructional design principles of Minimalist Instruction [1] rather than the standard systems approach [2]. With this newer approach the locus of control is with the student rather than with the system and students have greater freedom of exploration.

This paper discusses the computer hardware and software requirements for an ideal hypermedia system, aspects of current hypermedia authoring systems, the motivation and system overview for HyperGraph, the status of HyperGraph, the implementation of a sample module, and the development team.

## Hardware and Software Requirements

A hypermedia system requires substantial computer capability, but with the rapid pace of progress in computer hardware and software, appropriate systems will soon be available to educators. These will be capable of photorealistic image display, real-time graphics processing, sound generation, the ability to play stored video, and will have extensive hypermedia capabilities.

The ideal hardware capabilities include a processor capable of 50 to 100 MIPS and 10 to 20 Mflops, with a system main memory of 16 to 32 Mbytes. This is necessary not only for managing the hypermedia but also for accompanying simulations, such as the real time computation of images using different shading models. This computational power also would be required for hypermedia systems for learning chemistry or physics, where students might want to perform simulations, such as quantum mechanical computations for producing molecular orbitals or reaction surfaces.

The system also should have a large screen monitor (19 inch) with high resolution (1280 by 1024) 24-bit color graphics. Experiments have shown that an increased screen size improves the speed and comprehension of reading text from a computer screen [4]. A fast high capacity (500 to 1000 Mbytes) mass storage device is necessary to hold the hypermedia lessons, static images, and stored video. This might be a fixed disk drive, a CD-ROM, or a combination of these.

The software requirements include a hypermedia authoring system plus a good program development environment for the demonstration programs. There are several hypermedia products currently available for

IBM PC compatibles or the Apple Macintosh. These current systems vary in their capabilities and user interaction style.

## Current Hypermedia Authoring Systems

Current hypertext systems tend to be either card- or file-and-window based. The card-based systems, such as HyperCard for the Macintosh or Spinnaker Plus [5], for both the Macintosh and IBM PC, are limited in that only one card at a time can be displayed while the user might want to look at two or more different topics simultaneously. Another problem with card-based systems is that the card size is fixed and is usually given in terms of the number of horizontal and vertical pixels, say 640 by 480. The card can be larger than the actual screen and the extra material can be viewed by scrolling, but the entire card is stored whether there is information on it or not. If additional information is added to a card, it might overflow and a new card must be created.

Fixing the card size in the application hinders portability. For example, an application might be developed for a Macintosh with a small screen size. When this is moved to a Macintosh II or an IBM VGA system, the card no longer fills the screen. This would be worse if the application were moved to a workstation with a high resolution display. The card systems do have some advantages; for example, it is easier to define fixed areas for data base entries or links.

With file-and-window based systems such as Guide [3] for the Macintosh and IBM PC, one or more windows can be opened and scrolled to access all of a disk file of text and graphics. The windows can be dynamically resized and thus are easy to port to systems with different screen resolutions. A disadvantage is that it is easy to get lost with a large group of open windows. A computer science analogy for card-based versus file-and-window based systems is defining arrays versus creating dynamic data structures using pointers.

Another useful feature would be to have the hypertext links independent of the source material [6]. This would permit easy modification of the material, as in a word processor, without disturbing the link structure. It also provides the ability to have several different sets of links for the same material.

Current systems all lack important features but are being rapidly improved. For example, the IBM systems have poor color capability because they use the high-resolution (640 by 480) mode of the VGA, which has only sixteen colors. This will be improved soon, as they take advantage of more advanced graphics cards. An advantage of the Macintosh is the existence of inexpensive 24-bit color cards at 640 by 480 resolution, but few authoring systems can take advantage of them.

One requirement of the authoring systems is greater interoperability, the ability of the final system to run on different platforms. These systems will not be successful if they limit educators to a particular hardware vendor. This means that the systems should run on a Macintosh, IBM (under Windows or OS/2) and an X Windows UNIX workstation. There are currently some systems that run on both IBMs and Macintoshes, as noted above, but these don't run under the X Window System (some are being ported to X, as noted below). Commercial hypertext authoring systems are just being developed for UNIX workstations.

## *Motivation and System Overview*

As the field of computer graphics increases in scope, it becomes increasingly necessary to convey more information to students in courses. We need to improve radically the learning of current course material and to incorporate more and newer material into the courses. This extreme improvement will not occur as long as we rely on the traditional lecture and demonstration-based methods of instruction.

A unique feature of computer graphics is that courses bearing the same name, with similar but not identical material, might be taught in different academic areas, such as art, business, education, computer science, mathematics, engineering, and the sciences (where it might be called visualization). Computer graphics is a discipline that is ideally suited to hypermedia. Since the result of a graphics algorithm is an image, it is useful for the student to read a text description, study the detailed mathematical algorithm, view a pre-computed image, and then execute a program to generate variations of the images.

This would allow the students to determine the trade off of computational cost versus image quality for different algorithms, which is an important aspect of computer graphics. It also would allow the students to determine the effect on the computed image of varying the input parameters for a single algorithm. Since the generation of realistic images requires the synthesis of different techniques from different areas of computer graphics the ability to move easily from one subject to another is necessary. For example, as students study different shading algorithms they might want to study color theory and be able to switch easily between lessons on the two topics. Since computer animation is an important topic, the ability to view stored video is necessary.

Students frequently have difficulty understanding the relationship of different areas of a discipline, or why they must learn a particular topic so that later they will be prepared for other topics. To address this issue, the entire discipline will be presented as a three-dimensional, multi-level directed graph representing the three-dimensional information space containing computer graphics. Each node of the graph represents a sub-area in computer graphics, and the directed edges indicate the prerequisite nodes for that node and the nodes for which that node is a prerequisite. Thus, students could easily understand the position and context of a given topic within the entire field.

This overview should always be accessible to students, at varying levels of detail, so they can zoom out to see the relationship of the module they are currently studying to the entire field and then zoom in for a closer look at related modules. This overview is also helpful to the system developer. This requirement means that the hypertext authoring system should automatically construct this overview as the developer creates nodes and links. Commercial systems currently do not have this capability.

The graph nodes consist of subnodes that are hidden in the overview, but can be expanded to show the different sub-areas of a particular area. For example, rendering would be a top level node. This node has several subnodes, such as illumination models, which in turn consist of other sub-sub nodes, and so on. Thus, while an overview of the entire field of computer graphics can be displayed in the top view, an entire subfield can be viewed in a lower level view, and this hierarchical structure can continue as necessary.

Students should be able to choose the node expansion level of the system and thus the complexity of the overall view. They should be able to navigate visually through the space, traveling from node to node at any level of expansion, and should be able to rotate the space to obtain different views. Each bottom level node represents a unit of knowledge with an associated lesson. Each node can contain different types of lessons, targeted toward the different disciplines. For example, an art student might choose the associated art oriented lesson while a computer science student might choose the associated computer science lesson. However, the art student might want to look at the computer science lesson and vice versa for a different perspective on the same topic.

The system also will incorporate certain artificial intelligence techniques, particularly those pertaining to Intelligent Tutoring Systems (ITS). The workstations will be on a network with a central server. This server can compute and store a student model for each student. This model would be a record of the modules the student had tried, including those mastered and those not mastered, the student's performance on the module examinations, and an assessment of the student's knowledge of the subject. The system then could suggest review modules for the student.

Additionally the ITS can act as an advisory guide for students. In a complex three-dimensional information space the students may need help in navigating through the space. For example, students might want to study a particular application such as the generation of realistic images in CAD systems using models built from Constructive Solid Geometry methods. If students knew a sufficient amount about the topic then they could navigate through the three-dimensional information space to select all the correct nodes. If they were unsure, or perhaps wanted to save time, they might ask the system guide for help. The guide would carry on an interactive limited-domain natural-language dialogue with students until it was sure what they wanted. Then the guide could suggest a set of nodes to visit and study.

Students could combine into teams for group projects to produce and use larger graphics systems. These might be interdisciplinary teams consisting of scientists, artists, and computer science students. An appropriate project for a computer science student would be to build up a substantial three-dimensional graphics package in the course. The computer science version of the modules might each include some computer code implementing the subject material of that module, for example, a shading algorithm. This might be an object, assuming the students are using an object-oriented language. Successive modules might have different objects or additional methods for a previous object. These objects might not be complete, because the instructor might want the students to implement some algorithms themselves.

In a team project, all the computer science students might cover the same core material, but then different team members might cover different advanced modules and incorporate the different objects into a team graphics software system. While the computer science portion of the team was building the graphics software system, the artists on the team would be learning to use it in the construction of different types of images while the scientists would use it for visualization of scientific data. A team project such as this would be a good way to incorporate problem solving into the course since the team, consisting of many disciplines, might be given an overall problem and use the graphics techniques to solve the problem.

## *Status of HyperGraph*

We are currently implementing HyperGraph on an IBM PS/2 system using Guide 3.0 [3] running under Windows 3.0. We have chosen Guide for two reasons: it is a window based system and for its interoperability features. Versions of Guide run on the Macintosh and IBM PS/2, and a UNIX version is under development and should be available by the time this paper is published.

Guide supports four types of hypertext links. The first type is an annotation or "pop-up" link where the new text appears and remains only when the mouse button is clicked and held down. Annotation links are used only for small text sections. The second type is a replacement or expansion link. The replacement text (or graphic) appears when the mouse button is clicked and remains until the button is clicked again. The third type is a cross-reference or go-to link. Clicking the mouse button causes the cursor location to move to another area of the present document or to another document. If it is to another document then a window appears on the screen with the cursor positioned in that document. The fourth type of link launches an application. This can be used to display a previously created image or execute a program. The different types of links can be shown to the user by a change in the cursor appearance, different text styles or colors, or a combination of all three.

The initial development is being performed at Georgia State University (GSU) under the direction of the author. We plan to develop this initial prototype version and then to have the other developers on the team (discussed below) make their additions and modifications. The initial version is from the computer science viewpoint and makes use of materials the author has developed from teaching his courses and for the GSU-National Science Foundation (NSF Grant Number USE 8954402) Workshop on Computer Graphics at GSU in August, 1990. We have updated and improved these materials considerably since the workshop.

We have entered the equivalent of over two hundred pages of text, diagrams, and color images into Guide. All diagrams and images are stored in either TIFF or GIF format for portability. The current material covers the equivalent of a first course plus part of a second course in computer graphics. We have also developed some interactive demonstration programs. An

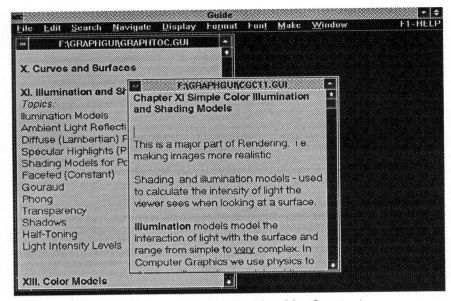

Figure 1: Part of HyperGraph's table of contents

example of the use of stored images and interactive programs is given below for a specific topic.

## *Design and Implementation of a Sample Module*

As an example of HyperGraph, we will discuss one currently implemented section: a chapter on the three-term Phong illumination model and shading models for geometrical objects represented as polygon meshes. This discussion shows some of the different types of hypertext links available in Guide by presenting figures that are snapshots of the screen.

Figure 1 shows part of the Table of Contents for HyperGraph. This will soon be converted to a graphical format, but initially we used text. The Chapter XI heading, "Illumination and Shading Models," has been expanded to show a detailed list of the contents. When unexpanded it appears just as the chapter heading above it, "X. Curves and Surfaces," with no additional text. The first item, "Topics," is a cross-reference link that has been activated, causing a second window to open on the screen with the cursor positioned at the beginning of that section. This second window can be moved and resized and the text automatically adapts to the different sizes. The following figures all show the windows expanded to full screen.

# Hypergraph – A Hypermedia System for Computer Graphics Education

The text portion of the module discusses the three contributions to the illumination model: diffuse reflection from ambient light and from a point light source, and specular reflection from a point light source (using the

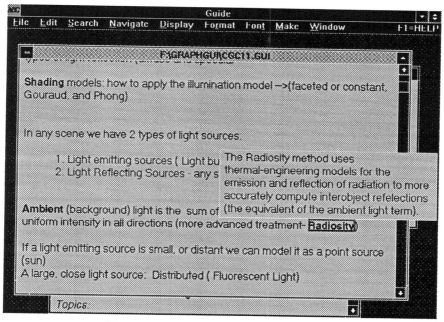

Figure 2: A pop-up link in HyperGraph

Phong $\cos^n$ approximation). Figure 2 shows a pop-up link that gives a short explanation of the radiosity method. This also will include a cross-reference link to the module treating radiosity in depth. The computer science view includes the algorithms necessary to construct the appropriate light and reflection vectors. Figure 3 shows two ways to construct the reflection vector. Each of these has an expansion link. If the derivation of these formulas is needed, then they can be shown by clicking on the appropriate expansion link, which is the bold faced text string, "1. Phong method. Assume light source is along Z-axis," and get a more complete explanation of the topic (Figure 4).

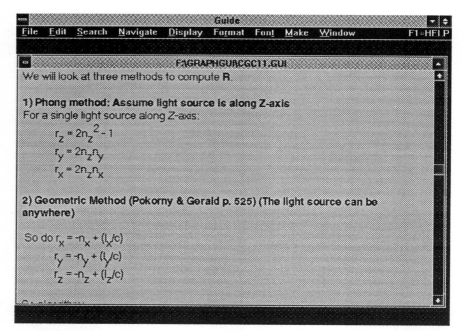

Figure 3: A simple explanation in HyperGraph that can be expanded

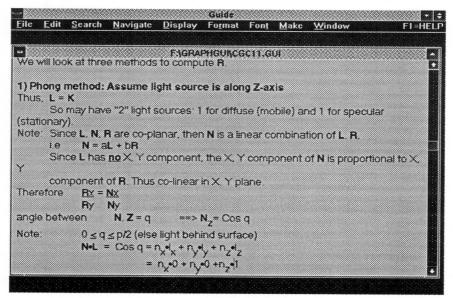

Figure 4: An expansion link in HyperGraph

The next section discusses three shading models that can be applied to objects composed of polygon meshes. The first is the faceted shading model in which one color intensity is computed and applied to the entire polygon. The second method is Gouraud shading in which a color intensity is computed for each vertex. The intensity at each pixel is then found by using bilinear interpolation of the vertex intensities. The third model is the Phong shading model in which the intensity at each pixel is computed from the normal at that pixel. The normal for each pixel is determined by bilinear interpolation of the vertex normals.

These three shading models are illustrated by stored images of the same scene shaded with the three models. Each image is displayed by clicking on a link in the discussion, so that the image representing Gouraud shading is displayed by clicking on a link that is the text string, "Display Gouraud shaded image." This link launches an application program that switches the VGA into the low-resolution graphics mode (320 by 200 with 256 colors) and then displays the image. When the student presses return, the mode is switched back to the high resolution graphics mode and the Guide screen reappears.

After the discussion on shading models there is another application link that can launch the program that created the stored images. This program allows students to change the viewing and lighting parameters and the shading model so that they can directly determine their effect upon the final image. By producing the images in real time students also can gain an appreciation for the image-quality versus computational-time tradeoff: the faceted and Gouraud images are computed and displayed rapidly, but the display of the Phong image is much slower.

Students can use different image data files with the program so they can display different images. This also allows students to determine the effects of the shading model on the required data structure. For example, a simple cube, if defined with six faces and eight vertices, loses its edges when either Gouraud or Phong shading is used. To maintain the edges a different data structure must be used where the vertices are each defined three times, once for each polygon.

The Gouraud shading model has difficulty with specular highlights. Since intensities are only computed at the vertices, if a specular highlight falls in the center of a polygon it is totally missed. Thus, in an animation sequence with a moving light source, the specular highlight might be at a vertex position, where it is visible, then move to the center of the polygon, where it would disappear, then move to another vertex, where it would reappear. For a simple scene, real-time animation could show this effect, or else a stored video segment could be used. For a sequence of this sort, an explanatory audio would accompany the animation. Real-time animation would be preferred so that the student could manipulate the parameters, such as the path of the light source and the number and size of the polygons in the polyhedron. This has not yet been implemented.

## Future System Development

As discussed above, the full development of HyperGraph will not be easy and will require expertise in different areas. The system development will be a team effort of the ACM SIGGRAPH Education Committee. The author, Chair of the Education Committee, is the Project Director. The other members of the development team (all are on the Education Committee) are:

- Jeff McConnell, Associate Professor of Computer Science at Canisius College, New York,
- Barbara Mones-Hattal, Assistant Professor of Art at George Mason University, Virginia, and
- Mike McGrath, Professor of Engineering at the Colorado School of Mines, Colorado.

We have begun to form an International Advisory Group (INAG) that will consist of Computer Graphics experts from both industry and academia representing the different disciplines. The INAG will review both the design and the implementation of the system as it is being developed. Its first members are Donna Cox, Steve Cunningham, and Steve Keith. Donna Cox, Director of the Renaissance Educational Computer Graphics Laboratory at the National Center for Supercomputing Applications, University of Illinois, Champaign-Urbana, combines expertise in art and scientific visualization. She has acquired international recognition as an authority on the application of artistic techniques to scientific visualization. Steve Cunningham, Professor of Computer Science at California State University Stanislaus, is the past chair of the ACM SIGGRAPH Education Committee and an internationally recognized expert in computer graphics education. Steve Keith works for Sterling Federal Systems, NASA Ames, and has had extensive industrial experience in computer graphics. He has also been heavily involved in ACM SIGGRAPH activities, especially educational activities.

## Conclusion

We have begun developing HyperGraph, a hypermedia system to teach computer graphics for artists, computer scientists, scientists and engineers. We are designing HyperGraph for easy enhancement and interoperability so that it will be able to execute on different platforms and incorporate the coming advances in computer hardware and software technology.

## References

[1] Carroll, J.M. *The Nurnberg Funnel: Designing Minimalist Instruction for Practical Computer Skill.* MIT Press, 1990
[2] Gagne, R.M., L.J. Briggs, and W. Wagner. *Principles of Instructional Design,* 3rd ed. Holt, Rinehart and Winston, 1988.
[3] Guide 3.0 is a product of OWL International, Inc., 2800 156th Ave., SE., Bellevue, WA 98007-9949.
[4] J. Nielsen. *HyperText & HyperMedia.* Academic Press, 1989.
[5] Spinnaker Plus is a product of Spinnaker Software Corporation, 201 Broadway, Cambridge, MA 02139
[6] Van Dam, A. "Electronic Books and Interactive Illustrations," In *Interactive Learning Through Visualization: The Impact of Computer Graphics on Education,* S. Cunningham and R. Hubbold, eds., Springer-Verlag, 1992.

# Hyper-Simulator Based Learning Environment to Enhance Human Understanding

### Shogo Nishida

Phenomena in large-scale systems have recently become more complex for operators, because systems have become larger and more complex. There is a strong need to develop a powerful tool which is useful for operators to deeply understand phenomena in large-scale systems.

However, software and hardware technologies in computer science are rapidly progressing, and it has become possible to make computers become thinking and communication tools based on new concepts [1, 2]. There is much research on computer-assisted instruction systems (CAI). A central concern of CAI is to make computers more intelligent by applying recently attained research results on cognitive science and artificial intelligence, such as knowledge representation and student modeling.

We believe that the most important approach to designing computer-supported systems for operators' understanding is by "putting intelligence in people's head, not in machines" [3].

In this paper we propose a concept of *hypersimulator* as a powerful computer-supported system to enhance operators' understanding of complex systems. Hypersimulator is literally a simulator fused with the concept of hypertext or hypermedia. The system's design approach is user centered [4]. First, as the preliminary investigation, the characteristics of operators' ways of thinking are analyzed and some notions of cognitive science are introduced. Secondly, the concept of hypersimulator is presented. Finally, the effectiveness of the system is demonstrated using simulation of some complex phenomena in power systems.

## Preliminary Investigations of Human Understanding

Without a richer and more detailed understanding of how people understand complex systems, computer-supported systems will be severely limited. To get a basic design concept of the system, we investigate characteristics of power system operators' ways of thinking and how cognitive science explains people's understanding with cognitive notions.

## *Characteristics of Operators' Ways of Thinking*

We investigate how operators understand the complex behaviors of power systems through interviewing operators. The following five characteristics are clarified:

1. Spatial and Causal Understanding: Topological information and spatial information of power systems, such as how loads and generators are geographically distributed in the systems, play an important role in operators' understanding of behaviors of power systems. Causal interpretation of operators also appears.

2. Qualitative Understanding: Qualitative information, such as direction of change of variables and relationships between the variables, plays a more important role in operators' understanding than quantitative information such as the magnitude.

3. Understanding by Analogy: As easy as conceiving from analogies between "power flow" and "water flow" which appeared in primary books of electric engineering, operators understand the behavior with images of power flow as water flow and power voltage as water pressure.

4. Understanding by Simplifying: It is difficult to understand the behavior of large power systems. Operators often simplify complex power systems and understand them using the simplified models. After that, they expand the simple systems gradually to more complex systems. In this understanding process, the kind of knowledge and concept used is changed according to abstraction level of power systems.

5. Situated Understanding: Ways of understanding depend on the situation, that is, what the operators are interested in. For example, voltage and reactive power are central concerns for operators in the voltage control process, while other factors such as current and active power are beyond their concerns. Situated understanding simplifies the behavior of power systems by filtering pieces of information and reducing them. However, when operators are captured by the situation, it becomes difficult to have other viewpoints of it.

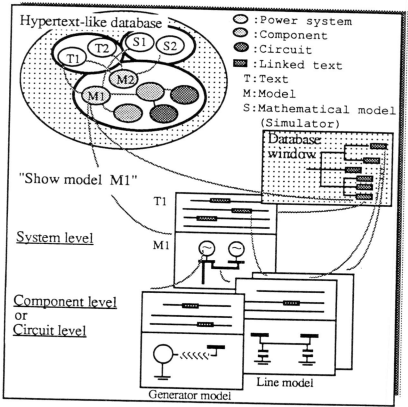

**Figure 4: Structure of database**

2. Database Structures: The database of SIMPLE is a hypertext-like database which has generally three types of data: model data, text data, and mathematical simulators. Each model in the database is linked according to hierarchical structures of power systems, kinds of phenomena, and aim of learning. The correspondence between windows and the database is illustrated in Figure 4.

# Examples of a Test Phenomenon in Power Systems

The effectiveness of SIMPLE is evaluated using the so-called unstable voltage phenomenon in power systems. A simple system which causes the phenomenon is shown in Figure 5.

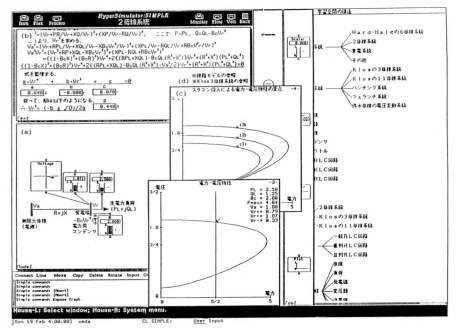

**Figure 5: Operator simulation screen**

The simulation model is shown on model pane (a) and explanations of the model are shown on text pane (b) in Japanese. The power versus voltage curves are drawn on view window (c) when a connected shunt capacitor is increased. Changes of some important variables and terms in equations which mathematically express the phenomenon are also shown on the text pane at that time. Therefore, the users can understand the phenomenon from multiple viewpoints, using theoretical, intuitive, and experiential understandings, while increasing the shunt capacitor by directly sliding the bar gauge of tool connected to it.

Text marked with a character '#' are linked to other simulation models. The users can easily access the linked models by clicking the text. For example, Japanese text (d) in Figure 5 means "See also Klos-Kerner three-bus system (which has more complex phenomenon)." Then the users can access the Klos-Kerner three-bus system by clicking this text and examine a more complex phenomenon.

The effectiveness of hypersimulator as a practical system is now under evaluation through operators' trial use. So far, we have received the following positive opinions from operators:

# Hyper-Simulator Based Learning Environment 87

- The hypertext-like database is powerful at organizing many models and phenomena in the ways each operator thinks.
- The direct manipulation interfaces make it easy to construct a new model and simulate it.
- Operators are expected to form effective mental models of power systems through the multiple views interfaces.

On the other hand, operators' opinions call for the further research on the combination with multimedia, such as video images and voice, and qualitative explanation.

## Conclusions

A hypersimulator, SIMPLE, is proposed as a thinking tool to enhance operators' understanding of complex phenomena in power systems. The effectiveness of the concept is confirmed using an example of a complex phenomenon in power systems. The features of SIMPLE are summarized as follows:

1. The proposed hypersimulator concept is useful for forming effective mental models of power systems in operators' minds and enhancing operators' deep understanding of complex phenomena in the power systems.

2. Connecting knowledge to action is achieved easily by putting the knowledge in the context of simulation. We believe that the learning style is pedagogically sound and effective for putting intelligence in operators' heads.

3. Object-oriented models adopted in the system are useful for developing complex user interfaces and software.

## References

[1] Conklin, J., "Hypertext: An Introduction and Survey," *Computer*, 20(1987), 17–41.
[2] Winograd, T. and F. Flores, *Understanding Computers and Cognition*. Ablex, 1986.
[3] Panel Session, "Intelligence in Interfaces," CHI+GI'87, Toronto, April, 1987.
[4] Norman, D. and S. Draper, *User Centered System Design*. LEA, 1986.
[5] Genter, D. and A. Stevens, *Mental Models*. LEA, 1983.
[6] Miyazaki, K. and N. Ueno, *Insights* (in Japanese), The University of Tokyo Press, 1985.
[7] Lampert, M., "Knowing, doing and teaching multiplication," Cognition and Instruction, 3(1987), 305.
[8] Miyake, N. (Ed: Y. Saeki), *What is Understanding?* (in Japanese), The Univ. of Tokyo Press, 1985.

# VISUAL THINKING
# AND
# VISUALIZATION

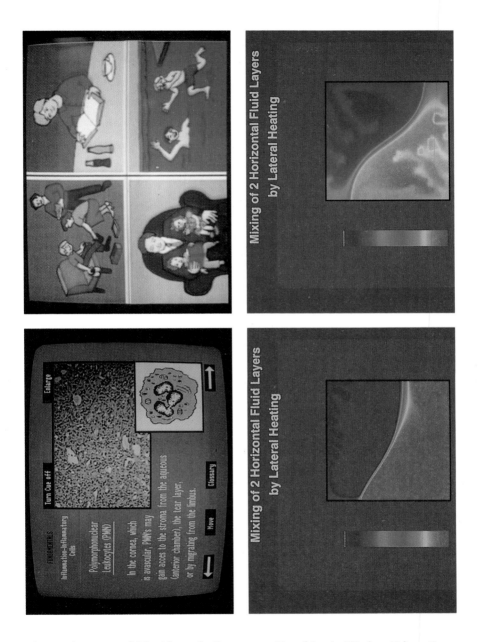

# The Multi-Faceted Blackboard: Computer Graphics in Higher Education
*Judith R. Brown*

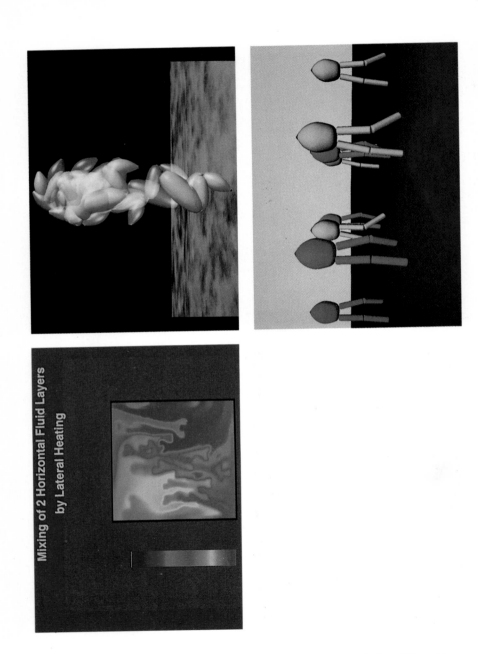

# The Multi-Faceted Blackboard: Computer Graphics in Higher Education
*Judith R. Brown*

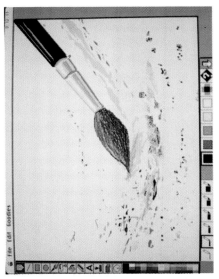

# Visual Ways of Knowing, Thinking, and Interacting
*Kenneth O'Connell*

above     Visualization of Concepts in Physics
*Hermann Haertel*

below     **Computer Assisted Lecturing: One Implementation**
*Jacques Raymond*

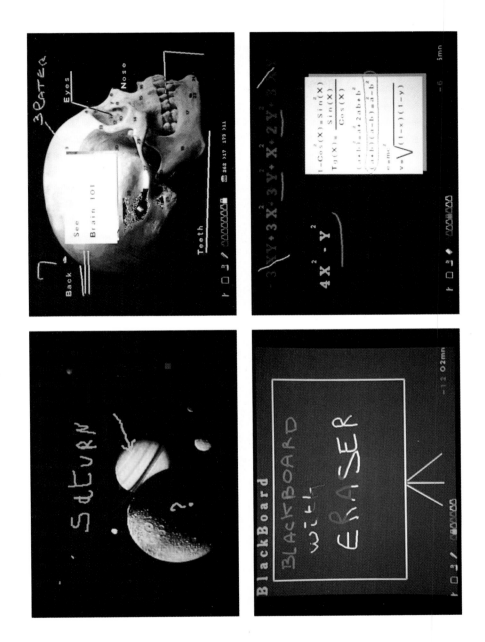

Computer Assisted Lecturing: One Implementation
*Jacques Raymond*

**Interactive Computer Graphics via Telecommunications**
*Joan Truckenbrod and Barbara Mones-Hattal*

**Interactive Computer Graphics via Telecommunications**
*Joan Truckenbrod and Barbara Mones-Hattal*

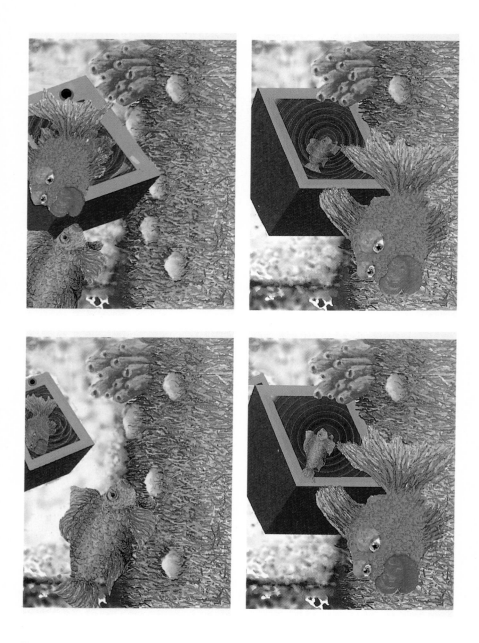

Interactive Computer Graphics via Telecommunications
*Joan Truckenbrod and Barbara Mones-Hattal*

**Interactive Computer Graphics via Telecommunications**
*Joan Truckenbrod and Barbara Mones-Hattal*

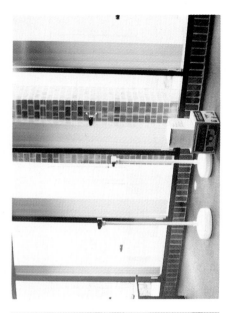

**Interactive Computer Graphics via Telecommunications**
*Joan Truckenbrod and Barbara Mones-Hattal*

Collaborative Computer Graphics Education
*Donna J. Cox*

Collaborative Computer Graphics Education
*Donna J. Cox*

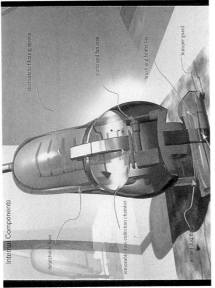

Collaboration between Industry and Academia –
Computer Graphics in Design Education
*Adele Newton*

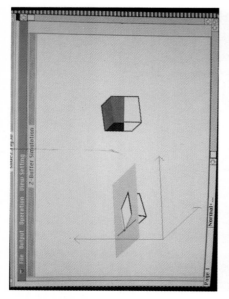
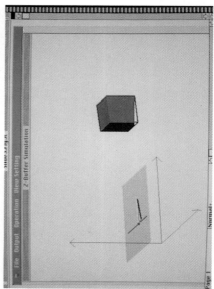
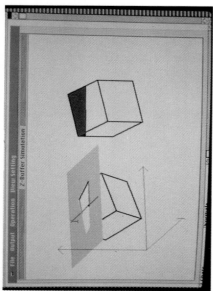

**Computer Graphics in Computer Graphics Education**
*Ahmad H. Nasri*

# Visual Thinkers, Mental Models and Computer Visualization

## Thomas G. West

### Graphic Computers as a Catalyst for Deep Change in Education

In recent years, computer graphics and scientific visualization have come to be seen as useful adjuncts to the conventional educational program in science, mathematics and other areas. However, this emerging family of technologies and techniques could be a catalyst for fundamental change in the educational process at the deepest levels, in time affecting what is taught as much as how it is taught.

From the earliest days computers were designed mainly to deal with numbers or letters. But in recent years even the least expensive and least sophisticated machines have come to be able to deal with graphics, in one form or another, with almost equal facility. With the coming of real graphic capabilities (whether interface or application, basic or advanced) an important change has occurred, although few may have appreciated it, partly because few have been trained or are accustomed to use effectively this fundamentally different and powerful medium.[1]

This situation, however, is not new. Before computers, of course, similar patterns were evident. Those who naturally gravitate toward a visual approach seem often to be in the minority, although they seem to be part of an especially creative and productive minority.[2] For example, the late nineteenth-century English statistician Karl Pearson and his son E. S. Pearson both relied heavily on visual images in their statistical work. They were surprised that their professional colleagues and students used visualization rarely, if at all. These associates believed that visualization was useful only for popular presentation, whereas the Pearsons saw visualization as the essential grounding for their most creative work.

In his lecture notes, Karl Pearson observed: "Contest of geometry and arithmetic as scientific tools in dealing with physical and and social phenomena. Erroneous opinion that geometry is only a means of popular representation; it is a fundamental method of investigating and analyzing statistical material."[3] Writing in the 1950s, the son lamented that "the prestige of mathematical procedures based on algebraic formulae is deeply entrenched in our lecture courses and our text-books, so that few mathematical statisticians will use to the full their visual faculties unless they are trained to do so."[4]

Similar difficulties are evident in more recent times. For example, visual-spatial approaches to teaching and investigating thermodynamics have been resurrected in recent years by some researchers. These researchers are using graphic computers to construct quantity surface plots for various substances, using methods that, about a hundred years ago, required the use of modeling clay and plaster of paris (or an unusually powerful visual imagination). This approach has been neglected for about a century but is ideally suited to powerful new graphic workstations. These researchers are finding this new approach to be rewarding in research and in education. But they are also encountering varied levels of resistance to this innovative approach among more traditional associates.[5]

In many areas within mathematics and the physical sciences, nonvisual approaches have held the dominant position for some time. However, with current computer graphic technologies, there is more and more evidence of major change. Some years ago articles were beginning to appear describing the new possibilities when computers were used to visualize statistical data. In one case, for example, it was noted that scientists at Stanford and at Harvard were "seeing patterns in data that never would have been picked up with standard statistical techniques," and that the intent of data analysis techniques "is to discover patterns, to find non-random clusters of data points. Traditionally, this is done by using mathematical formulas." However, with the coming of "computer motion graphics, it has become possible to look at three-dimensional projections of the data and to make use of the uniquely human ability to recognize meaningful patterns in the data."[6] In recent years, there is more evidence that visual approaches are moving back into center stage in mathematics and in certain related new fields such as "system dynamics" or "chaos."[7]

Viewing the increase of such changes, it would appear that we could be entering a phase of really fundamental change. Of course, we hear much about fundamental or revolutionary change. We have learned to be cautious.

But it may be argued that such terms could turn out to be appropriate now because we may be effectively tapping into basic neurological capacities, capacities that have long been relatively underdeveloped by formal education and common experience in modern industrial culture.

Conventional educational systems, at all levels and in most fields, have been heavily weighted toward verbal and numerical approaches to acquiring and assessing information. For example, observers have noted recently that "despite the obvious importance of visual images in human cognitive activities, visual representation remains a second-class citizen in both the theory and practice of mathematics."[8] Because of the heavy emphasis on the verbal, the symbolic, and the numerical, the visually-oriented methods of individuals such as the Pearsons have been more the exception than the rule. Especially in the last century or so, visual approaches to knowledge have been increasingly denigrated. For many, this situation is so deeply rooted that they are not even aware of an imbalance. The prestige of the nonvisual approaches has spread well beyond traditional academic settings into all sectors of modern economies, in many cases thoroughly eclipsing earlier methods of learning by imitation and experience and apprenticeship.

The pattern we might wish to consider for the future, however, is a major reversal of this trend. We need to see that it is possible to go forward by going backward, that is, by returning to earlier methods to help us move ahead. Once, the typewritten letter carried the most weight; more recently, the handwritten note commanded the most attention; now, most recently, the brief and informal electronic mail message. Similarly, in the future, when the situation is reversed once again — when the verbal-academic approach has spread and the new economy requires some very different skills — we might find that new forms of apprenticeship could become among the most desirable and effective routes to the most sought-after and productive positions.[9]

Our great, great grandfathers, whether craftsmen, sailors, merchants, or farmers, learned mostly from doing, from imitating and helping, from making errors, relying more on hand and eye than on word and phrase. Our fathers and grandfathers learned primarily from lectures and books; and in many cases they were not allowed into the world of adult work until quite late, and they rarely received further education.

Before long, however, we might see another major change in education and work. Before long, our sons and daughters may learn mostly from experience once again—actively, using all their senses—but this time they would use visual and kinesthetic computer simulations of reality as much as reality itself. And in this reversed new world, where education would alternate with productive work throughout life, the creative visual thinkers and imaginative experimenters may, once again, find themselves better adapted to changed circumstances than those who are mainly best at reading, retaining, and accurately recalling thousands and thousands of words.

## Visual Talents and Verbal Difficulties

There is evidence that some highly original and productive thinkers have clearly preferred visual over verbal modes of thought for many tasks. Historically, it is apparent that some of the most original thinkers in the physical sciences, engineering, mathematics, and other areas relied heavily on visual modes of thought, employing images instead of words or numbers.[10] However, some of these same thinkers have shown evidence of a striking range of verbally-related learning problems, including difficulties with reading, spelling, writing, calculation, speaking, and memory. What is of greatest interest here is not the difficulties themselves but their frequent association with marked visual and spatial talents.

These associations seem not to be accidental. Several neurologists believe that sometimes highly visual talents are closely associated with various forms of verbal difficulties because of certain early patterns of neurological development. Recent research suggests that some forms of early brain growth tend to suppress the development of the more verbal left hemisphere and promote the development of the more visual right hemisphere, producing an unusual symmetry of brain form and function.[11]

Varied visual talents mixed with verbal difficulties are evident in a diverse group of important historical persons: Michael Faraday, James Clerk Maxwell, Albert Einstein, Lewis Carroll (Charles Dodgson), Henri Poincaré, Thomas Edison, Leonardo da Vinci, Winston S. Churchill, Gen. George S. Patton, and William Butler Yeats.[12]

In the life of Albert Einstein, for example, the importance of visual talents in conjunction with verbal difficulties is fairly well recognized. His poor memory for words and texts made him hate the rote learning methods of his earlier schooling. However, he tended to thrive later at the progressive school that was based largely on the visually-oriented educational principles of Johann Heinrich Pestalozzi.[13]

The coexistence in Einstein of visual talents along with verbal difficulties has been noted by several observers. The physicist and historian of science Gerald Holton has remarked that "an apparent defect in a particular person may merely indicate an imbalance of our normal expectations. A noted deficiency should alert us to look for a proficiency of a different kind in the exceptional person. The late use of language in childhood, the difficulty in learning foreign languages ... may indicate a polarization or displacement in some of the skill from the verbal to another area. That other, enhanced area is without a doubt, in Einstein's case, an extraordinary kind of visual imagery that penetrates his very thought processes."[14]

When one follows Einstein's visual-spatial propensities to those he most admired as a young man, such as Michael Faraday and James Clerk Maxwell, one finds not only a pronounced visual approach, but also varied verbal difficulties as well.[15]

## Visual Talents with New Visualization Tools

It is expected that recent developments in the use of graphics-oriented personal computers and in scientific visualization using graphics workstations and high-speed supercomputers could be an initial phase in an increasingly significant shift in our information economy. This shift is seen as moving toward a greater emphasis on visual approaches to the analysis of complex information and away from an excessive emphasis on words and numbers alone. Such a shift could greatly benefit highly creative visual thinkers, because, in time, these individuals could prove to be among those best suited to deal with the major new directions of development promoted by this shift.

Some students with special visual proficiencies, however, if the neurologists are correct, may require certain adjustments in educational expectations. For example, there is evidence that the maturational lags often seen in "late bloomers" and some visual thinkers can eventually yield higher levels of neurological and intellectual capacity than is usually seen among those who mature rapidly.[16] Thus, some of those students who have difficulty in early schooling (those for whom the "easy" is hard) can sometimes do surprisingly well at higher levels of education (when the "hard" becomes easy) and may be far more creative and productive in later life than those with "ordinary" brains, however bright and well organized. On the other hand, some bright students, who are well focused and well organized and have no difficulty with memorized material, may turn out to be especially ill-suited to do truly advanced and original conceptual work, especially when these tasks draw mainly on highly developed visual-spatial capacities.

There is also evidence that the early neurological growth patterns associated with visual abilities and mixed talents can be generators of great diversity in brain structure and function.[17] Many of those affected can be expected to show a wide range of special talents as well as highly heterogeneous forms of learning difficulties. This aspect makes these individuals difficult to identify since their mixed talents and difficulties occur in many different combinations. However, if properly understood and developed, this great diversity may be profoundly beneficial to specific institutions and corporations as well as the larger society.

## Graphic Computers Could Promote a Shift from Mass Lectures to Personal Tutors

Graphic computers may be instrumental in reversing another long-term trend: allowing each student to have his own personal tutor for some substantial part of their time, rather than relying mainly on traditional mass education in classrooms and lecture theaters. Such a change could be seen as a major reversal, a departure from nineteenth-century methods of mass education and a return to the educational methods common among privileged groups in ancient times and the Middle Ages.

In the not-too-distant future, substantial use of computer "tutors" could free teachers from more repetitive presentations and allow for more individual attention and higher-quality group interaction, as in working groups and seminars. The teacher is not replaced but made more effective. The printed book is not replaced but integrated with a learning medium that is more interactive and responsive. And throughout, it is expected that the power of moving visual material would be employed wherever possible.

This perspective presumes, of course, continuing reductions in hardware and software costs as power, capability, and ease-of-use increase. Although some consider such systems as comparatively expensive at this time, this perspective also presumes that, in the near future, such machines (whether personal computers or workstations), when used properly, will be comparatively less expensive than human labor alone. The potential of such computer "tutors" is further extended by ancillary developments such as inexpensive mass storage media, interactive video, multimedia, simulators, artificial reality, and other related innovations on the near horizon. When properly designed and used, such technologies should expand the scope of possibilities while overall costs are reduced.

## Recent Convergence of Trends toward Visual Approaches

In the early 1990s a number of trends appear to be converging to promote a gradual but dramatic change in education, along with corresponding changes in the roles and methods of knowledge workers in the larger economy. Several interrelated trends appear to be significant.

Some of the problems now seen to be of greatest importance, such as large atmospheric or ecological systems, are by their nature problems of vast, inherent complexity. Complex and voluminous data from satellites and other automated sources has tended to overwhelm traditional methods of analysis.[18] Graphics-oriented computers, as they increasingly become more powerful and less expensive, promise to be an effective tool for original work in many fields, partly because such systems are inherently better suited to dealing with such complex phenomena by focusing on the whole rather than the parts.[19]

This trend is supported by the recent emergence of new methods and techniques for the analysis of complex systems, such as chaos theory and dynamical systems studies. The raise of interest in fractals and related visual approaches, serves to heighten awareness of the importance of these developments.[20] In addition, some contemporary mathematicians are coming to see their discipline less as a matter of symbol manipulation and logical rigor and more as a science of patterns, shifting their emphasis toward visual and intuitive approaches.[21]

As previously noted, many fields of mathematics and physical science have long accorded high status to spare and rigorous symbol manipulation; diagrams and pictures have had low status. However, some highly creative visual thinkers have always had a passion for geometry and other visual-spatial approaches to mathematics. More recently, some advanced mathematics research has returned to an interest in geometry. For example, one layperson's breakthrough in the mathematics of tiles ("combinatorics") used direct visual-spatial methods that had previously been considered unacceptable by professional mathematicians but have since become widespread.[22]

Another significant trend is the growing awareness that some who are very talented visually may perform rather poorly in conventional educational systems. Recently, a professional journal for electrical and electronic engineers has carried an opinion article arguing that engineering schools are screening out some of the best future engineers because they are mild or compensated dyslexics. The article argues that some of the finest engineers are very visual in their way of working but often have difficulties with words or numbers. These difficulties inevitably result in unimpressive academic performance. As a consequence, it is argued that some of those who could become the most talented engineers either drop out of increasingly demanding academic programs or fail to move into the most important job positions because of mediocre grades. Since the article appeared in December 1990, a number of letters to the editor have indicated that this pattern may be more prevalent than previously expected. One Ph.D. electronic engineer at a prominent national laboratory thanked the article author for "letting the bogeyman out of the closet," observing that it had been "a shameful secret until now." Another engineer responded: "I can emphatically agree ... 100% valid. ... Regrettably, I didn't come to this realization until after ... graduate school." [23]

## From Medieval Clerk to Renaissance Man

For some four hundred or five hundred years we have had our schools teaching basically the skills of a Medieval clerk: reading, writing, counting, and memorizing texts. But with the deepening influence of computers of all kinds, it now seems that we might be on the verge of a new era when we will be required to cultivate broadly a very different set of skills: the skills of a Renaissance man such as Leonardo da Vinci.

As part of this change, in the not-too-distant future, past ideas of desirable traits could be gradually but dramatically transformed. In place of the qualities desired in a well-trained clerk, we might, instead, find preferable a facility with visual content and modes of analysis instead of mainly verbal (or numerical or symbolic) fluency; a habit of innovation through making connections among many diverse fields; the more integrated perspective of the global generalist rather than the increasingly narrow specialist; a propensity to learn directly through experience (or simulated experience) rather than primarily from lectures and books; a life-long habit of continuous learning in many different areas of study (perhaps with occa-

# Visual Thinkers, Mental Models and Computer Visualization

sional but transient specialization); an ability to rapidly progress through many stages of research and development and design using imagination and mental models along with three-dimensional computer-aided design.[24]

Leonardo da Vinci's emphasis on analysis through visualization may come to serve us as well as it served him, providing results well in advance of those who follow other more conventional approaches. It is interesting to note that some current advocates of three-dimensional modeling workstations are now speaking in terms almost identical to those used by Nikola Tesla, the extremely gifted visualizer who claimed to have performed all phases of design and testing in his head with his own mental models, explaining that it was simply wasteful of time and money to work in any other way.[25]

In the near future, it seems that we might be in a position to come full circle, using the most advanced technologies and techniques to draw on the most old fashioned approaches and capacities: to simulate reality rather than describe it in words or numbers. To learn, once again, by doing rather than by reading. To learn, once again, by seeing and experimenting rather than by following memorized algorithms. Sometimes the oldest pathways can be the best guides into uncharted territory.

There may be little choice about the coming changes in education and work. If we continue to educate people who have primarily the skills and perspectives of the Medieval clerk (no matter how advanced, specialized, or esoteric their field of study), we may increasingly be turning out people who, like the unskilled laborer of the last century, will be unable to compete with increasingly intelligent machines and will have less and less to contribute to the real needs of our culture and will have less and less to sell in the marketplace.[26] As one of the pioneers of information and control theory, Norbert Weiner, pointed out in 1948, the first industrial revolution, in producing goods with mechanical power, resulted in the "devaluation of the human arm by the competition of machinery." Similarly, he predicted that the second industrial revolution, involving computers in their many forms, is "bound to devalue the human brain, at least in its simpler and more routine decisions." Although some specialist workers will always be needed in specific areas, he explained that, on the whole, "taking the second revolution as accomplished, the average human being of mediocre attainments or less has nothing to sell that is worth anyone's money to buy."[27]

Some forty years later, we can see clearly that some time in the not-too-distant future, machines will be the best clerks. Given this situation, we must learn, as teachers and workers, to maximize in ourselves and in our students what is most valued among human capabilities and what machines cannot do. It seems clear that, increasingly, many of these most valued skills will involve the insightful and broadly integrative capabilities often associated with visual modes of thought, skills that can perhaps be taught and exercised most effectively using graphics-oriented computers.

## Notes

1. Jeff Johnson, Teresa L. Roberts, et al,"The Xerox Star: A Retrospective," in *Computer*, Vol. 22, No. 9, 1989, pp. 11–29.
2. It should be noted that broad definitions of "verbal" and "visual" (and, closely related, "spatial") capabilities are used here, referring in part to the varied but apparently antithetical thinking styles generally believed to be embodied in the left and right cerebral hemispheres, respectively. One psychologist, Howard Gardner, distinguishes several major forms of intelligence, but argues for the special status of visual-spatial intelligence in contrast to verbal intelligence: "In the view of many, spatial intelligence is the 'other intelligence' — the one that should be arrayed against, and be considered equal in importance to, 'linguistic intelligence.' 'Although Gardner does not subscribe to the dichotomization of intelligence into separate hemispheres, he says: "Still, I would admit that, for most of the tasks used by experimental psychologists, linguistic and spatial intelligences provide the principle sources of storage and solution." Howard Gardner, *Frames of Mind: The Theory of Multiple Intelligences*, Basic Books, New York, 1983, p. 177.
3. Karl Pearson, quoted in E. S. Pearson, "Some Aspects of the Geometry of Statistics: The Use of Visual Presentation in Understanding the Theory and Application of Mathematical Statistics" in *The Selected Papers of E.S. Pearson*. Presented as the Inaugural Address of the President to the Royal Statistical Society in 1956. University of California Press, Los Angeles, CA, 1966, p. 252.
4. E. S. Pearson, "Statistics," in *Papers*, 1966, p. 253.
5. Kenneth R. Jolls,"Understanding Thermodynamics through Interactive Computer Graphics," in *Chemical Engineering Progress*, February 1989, pp. 64–69; Kenneth R. Jolls and Daniel C. Coy, "The Art of Thermodynamics," in *IRIS Universe: The Magazine of Visual Processing*, No. 12, spring, 1990, pp. 31–36.
6. Gina Kolata, "Computer Graphics Comes to Statistics," in *Science*, Vol. 217, 1982, pp.919–920.
7. Lynn Arthur Steen, "Mathematics Education: A Predictor of Scientific Competitiveness," in *Science*, Vol. 237, 1987, pp. 251–252, 302. Lynn Arthur Steen,"The Science of Patterns," in *Science*, Vol. 240, 1988, pp. 611–616. Walter Zimmermann and Steve Cunningham, eds.,*Visualization in Teaching and Learning Mathematics*, The Mathematical Association of America, Washington, D.C., 1990. James Gleick, *Chaos: Making a New Science*, Viking, New York, 1987. Ralph H. Abraham and Christopher D. Shaw, *Dynamics — The Geometry of Behavior, Part 1: Periodic Behavior* and *Dynamics — The Geometry of Behavior, Part 4: Bifurcation Behavior*. The Visual Mathematics Library, VisMath Volumes 1–4, Aerial Press, Santa Cruz, CA, 1984–1988.
8. Jon Barwise and John Etchemendy, "Visual Information and Valid Reasoning," in Zimmermann and Cunningham, eds.,*Visualization*, 1990, p. 9. The resistance to visualization in

mathematics can run very deep. Mathematics would seem to be such a pure and abstract discipline that it would be relatively immune to fashion and the pressures of power politics. However, it is interesting to note that in one case, a secret society was formed in order to purify and reform mathematics mainly by doing away with all visual and diagrammatic content. Founded in France, this group of mathematicians, the Séminaire Bourbaki, was extremely influential in the early and middle parts of the twentieth century, forming an orthodoxy inhospitable to the approach of visually-oriented mathematicians such as Henri Poincaré. (Gleick, *Chaos*, 1987, p. 90.) In more recent years, the Bourbaki group itself seems to have softened, for example, inviting a speech by a visually-oriented mathematician whose work was inspired by children's Lego blocks. (Ivan Rival, "Picture Puzzling: Mathematicians Are Rediscovering the Power of Pictorial Reasoning," in *The Sciences*, January/February 1987, p. 46.) However, it seems that elsewhere opposition to the visual approach is breaking out once again among some mainstream mathematicians — those who are annoyed by the great success of Benoit Mandelbrot's "fractals" in recent years; one critic has argued that " 'fractal geometry ... has not solved any problems. It is not even clear that it has created any new ones.' " (S. Krantz, quoted in R. Pool, "Fractal Fracas: The Math Community is in a Flap Over the Question of Whether Fractals Are Just Pretty Pictures—Or More Substantial Tools," in *Science*, Vol. 249, 1990, pp. 363–364.)
9. Some of the material included here and elsewhere in this paper has appeared previously in different form in Thomas G. West, *In the Mind's Eye: Visual Thinkers, Gifted People with Learning Difficulties, Computer Images and the Ironies of Creativity*, Prometheus Books, Buffalo, NY, 1991, pp. 56, 305, et al.
10. On one level such a preference would appear to be obvious: that, say, a mechanical engineer would prefer a visual-spatial (right-hemisphere) approach in some aspects of his work. However, once one looks into the analytic training, one sees how the numerical and verbal (left hemisphere) approach often comes to dominate.
11. Norman Geschwind and Albert Galaburda, *Cerebral Lateralization: Biological Mechanisms, Associations, and Pathology*, The MIT Press, Cambridge, 1987; N. Geschwind and W. Levitsky, "Human Brain: Left-Right Brain Asymmetries in Temporal Speech Region," in *Science*, Vol. 161, 1969, pp. 186–187; N. Geschwind, "Why Orton Was Right," *The Annals of Dyslexia*, Orton Dyslexia Society Reprint No. 98, Vol. 32, 1982; N. Geschwind and P. Behan, "Left-Handedness: Association with Immune Disease, Migraine, and Developmental Learning Disorder," in *Proceedings of the National Academy of Sciences*, Vol. 79, 1982, pp. 5097–5100.
12. West, *Mind's Eye*, 1991, pp. 29–40, 101–175.
13. Gerald Holton, "On Trying to Understand Scientific Genius," in *The American Scholar*, Vol. 41, No. 1, 1971–1972, pp. 104–106.
14. Holton, "Genius," in *Scholar*, 1972, p. 102.
15. Faraday had minor speech difficulties, but major difficulties with memory and mathematical symbols. Maxwell was a stutterer all of his life and had difficulty organizing his verbal thoughts under pressure; however, Maxwell was known to write and speak with great clarity when given time to prepare his verbal material. (West, *Mind's Eye*, 1991, pp. 29–34, 101–118.)
16. Martha Bridge Denckla, "Motor Coordination in Dyslexic Children: The Theoretical and Clinical Implications," in *Dyslexia: A Neuroscientific Approach to Clinical Evaluation*, Frank H. Duffy and Norman Geschwind, eds., Little, Brown, Boston, 1985, p. 194. Albert

Galaburda, "Research Update," ODS Tape No. 30, Orton Dyslexia Society, Baltimore, MD, 1984.
17. Geschwind and Galaburda, *Cerebral Lateralization*, 1987.
18. Evelyn Richards, "The Data Deluge: Exotic Electronic Systems May Hold Key to Future Success," in *Washington Post*, Sept. 24, 1989, pp. H1ff; Richard Wolkomir, "NASA's Data Deluge," *Air & Space*, Vol. 4, No. 4, 1989, pp. 78–82.
19. Thomas A. DeFanti, Maxine D. Brown and Bruce H. McCormick, "Visualization: Expanding Scientific and Engineering Research Opportunities," in *Computer*, Vol. 22, No. 8, 1989, pp. 12–25.
20. Abraham and Shaw, *Dynamics*, 1984–1988.
21. Steen, "Patterns," in *Science*, 1988, pp. 611–616.
22. Rival, "Picture Puzzling," in *The Sciences*, 1988, pp. 40–46.
23. Walter Frey, "Schools Miss Out on Dyslexic Engineers," *IEEE Spectrum*, Vol. 27, No. 12, 1990, p. 6; Walter Frey, personal communication, March 5, 1991.
24. Peter R. Ritchie-Calder, *Leonardo & the Age of the Eye*, Simon and Schuster, New York, 1970. See also: Giuseppe Satori, "Leonardo Da Vinci, Omo Sanza Lettere: A Case of Surface Dysgraphia?" In *Cognitive Neuropsychology*, Vol. 4, No. 1, 1987, pp. 1–10 and P. G. Aaron, Scott Phillips and Steen Larson, "Specific Learning Disability in Historically Famous Persons," *Journal of Learning Disabilities*, Vol. 21, No. 9, 1988, pp. 523–545.
25. Compare the descriptions provided in the following: Nikola Tesla, *My Inventions: The Autobiography of Nikola Tesla*, Ben Johnson, ed., Hart Brothers, Williston, VT, 1919 and 1982, pp. 31–33 and KPMG Peat Marwick, *Competitive Benefits from 3D Computing: A Study of Silicon Graphics' Customers*, Silicon Graphics, Mountain View, CA, 1989, p. 10.
26. Hans Moravec, *Mind Children: The Future of Robot and Human Intelligence*, Harvard University Press, Cambridge, 1988; Hans Moravec, "Human Culture: A Genetic Takeover Underway," in *Artificial Life: Proceedings of an Interdisciplinary Workshop on the Synthesis and Simulation of Living Systems*, Christopher G. Langton, ed., Center for Nonlinear Studies, Los Alamos National Laboratory, Vol. 6, Studies in the Sciences of Complexity, Santa Fe Institute, Addison-Wesley, Redwood City, CA, 1989.
27. Norbert Weiner, *Cybernetics: Or Control and Communication in the Animal and the Machine*, the MIT Press, Cambridge, MA, 1948 and 1961, pp. 27–28.

# The Multi-Faceted Blackboard: Computer Graphics in Higher Education

## Judith R. Brown

A university environment is a mosaic of disciplines, graduate and undergraduate studies, and research activities. Through the nurturing of all these diverse elements, and the informational sharing and cross disciplinary activities made possible by modern technology, the individual pebbles in the mosaic form a vibrant and exciting, all-encompassing educational environment. Yet, they do so without losing their own individuality.

Computer graphics can serve these individual needs in individual ways, while also providing overall quality. While this is undoubtedly true about education at all levels, this paper addresses only the university environment. Computer graphics in higher education is used

- to gain insight through interaction with, and visualization of, data and simulations,
- to provide information to others through presentation materials, and
- as a medium for creative and artistic expression.

### EVOLUTION

Educational computing has changed tremendously in the past ten years, especially in the last five years, and there is every reason to expect this progress to continue over the next five years. Within the last ten years, campus computing changed from batch to interactive programs; university computing centers expanded from large centralized computing resources to include additional smaller distributed resources; and computing spread beyond the sciences into the arts and humanities. Within the past five years, we have seen computing power which was only available at large central facilities become available in a desktop size and at an affordable price for a university departmental lab. We have seen educational software change from routine drill and practice applications to exciting, interactive simulations. And, we now see students coming to campus with their own computers and a high degree of computer skills.

*FUTURE*

In the next five years, we can expect fast, high-resolution graphics workstations, which are now finding their way into research labs, to be available in student labs and classrooms. Every campus will need high speed networks in departments, among departments, and into dormitories to enable rapid flow of information and access to central or departmental facilities which provide high quality hard copy and video. The use of animation, both real-time animations on workstations and the production of video, will become increasingly important.

*SCENARIOS*

The following scenarios from Second Look Computing at The University of Iowa serve as models for using technology for education and served as the impetus for a multimedia facility for faculty. For more information on this facility, see [5].

### *Brooks Landon: Is it Real or is it Memex? A New World*

In 1945 Vannevar Bush's article "As We May Think" appeared in the Atlantic Monthly. That article has been lifted from obscurity in recent years because it was the first statement with a vision of universal access to all formats of information — text, audio, and video — that he called "Memex" and that we generally categorize today as "hypermedia." While the trivial nuts and bolts of the hardware differ from Mr. Bush's vision, the conceptual basis of today's state-of-the-art information systems derives quite purely from this landmark piece.

Brooks Landon, a professor in the Department of English, wants to create a "hypermediable" teaching environment: a computer based, high-fidelity audio and video arena for faculty in English and Philosophy. Brooks is teaching a course on twentieth century culture and writing. He has amassed a large number of images related to this course as well as many movies from the period. As his students see early decades recreated with high fidelity, they will have a better understanding of that period.

What does Brooks want? He wants a room with a Macintosh II, videodisc player, and videotape unit all connected to a video projector, with a high-fidelity, sound-surround audio system. From it he can deliver — in a multimodal form — the sights and sounds, the critical events and thoughts, that shaped our century. And so can his students when they complete interactive programs rather than twenty-page term papers.

## *Joanne Eland: Taking the Pain Out of Learning*

Jo Eland is a professor in the College of Nursing. She is a nationally-known specialist in pain research with an emphasis on children's pain. After attending the first short course on HyperCard offered at the Weeg Computing Center, she decided that HyperCard would be an ideal environment to set up simulated patient care scenarios for student exploration and learning. Each HyperCard stack was designed to to provide "real world" patient experience based on a theoretical concept covered in the didactic portion of the acute care clinical course. She has created and published twenty-seven of these stacks with integrated graphics and sounds for use in the nursing community.

An important feature of the stacks is the opportunity they provide for students to practice clinical decision making. Dr. Eland programmed the stacks to provide situations where students must make clinical decisions about a patient's care. Feedback is provided regarding the accuracy of decision and rationale for inappropriate choices. This feedback makes these packages unique. Most traditional instructional approaches in nursing education cannot provide students with an opportunity to practice clinical decision making, the foundation of professional nursing practice. Such skills had to be developed in practicum settings where opportunities to practice problem solving cannot be systematically structured for each student, and instructor feedback is inconsistent at best. The computerized patient simulations that Dr. Eland has designed provide an arena in which analytical skills can be consistently practiced by all students without risk to patient safety.

Dr. Eland believes that the Macintosh and HyperCard are a proven environment that can enhance the learning experience and provide learning opportunities that did not previously exist. With the additional ability of HyperCard to drive external devices such as a videodisc player or a slide projector, she feels that it will be possible to produce instructional stacks which will have far greater impact than those she has previously created.

## Robert Folberg: Pathology of the Eye

Robert Folberg, associate professor of Ophthalmology and Pathology, developed Pathology of the Eye, a complete interactive course covering thirteen topics in ocular pathology. It is a course for ophthalmology residents, advanced medical students, and for continuing education for ophthalmic practitioners. It is intended to permit the user to understand the natural history of diseases, the pathophysiology of ocular disease, and the pathologic basis for the appearance and treatment of ocular disease. Color Plate 1 shows a computer screen from Pathology of the Eye. There are versions of this software for both IBM PS/2 and Apple Macintosh II computers.

## Chris Roy: Look at These Masks, Listen to Those Drums: Sharing Scholarship

Chris Roy is a professor in the School of Art and Art History. His specialty is the art of the tribal people of Africa and the southern Pacific Ocean. He and his students have traveled extensively in these areas, and they have accumulated an impressive collection of slides, photographs, film, video, and audio recordings. Chris has produced videotapes from his original material which are being used by a number of institutions in their African art curriculum. He and Julie Hausman, curator of the Office of Visual Materials, are in the final stages of producing a videodisc containing images of objects from the Stanley Collection of African art at the University of Iowa Museum of Art. Information describing each of the images is held in a computer database which is keyed to its videodisc frame. The videodisc and database are used with a retrieval system running on a Macintosh. The system can display multiple digitized images together on a single monitor, allowing comparison of several objects at a time. He believes the system has the potential to make a major impact on the field of art history.

## Nancy Tye-Murray: Teaching Lip Reading

Nancy Tye-Murray, Associate Research Scientist in Otolaryngology, uses interactive computer graphics techniques to teach lip reading skills to hearing-impaired children. The child watches the instructor's lips on videodisc segments, makes selections on the computer, and is both assisted by the computer and rewarded for correct choices by computer graphics. The interface to the computer program is a touch screen, and the use of computer graphics is very important in communicating with the child. Color Plate 2 shows a computer screen with some of the options from which the child can select.

## *Students: Term (Papers) of Endearment*

Form should follow function, said architect Frank Lloyd Wright regarding buildings. The same holds true for knowledge: its form should follow its function. In the other scenarios, we focus on use by faculty and staff. Yet the University of Iowa exists primarily to educate the 29,000 students enrolled here. They are, and must remain, central.

The traditional assignment for a student is the typed term paper. However, many students are taking courses in which the knowledge they need to demonstrate is neither linear nor textual. In the visual arts, comparing two artists is best done by seeing examples of their works. In the performing arts, an understanding of music must relate to hearing the music itself. In the health sciences, high fidelity illustrations of the variety of ways in which diseases present themselves are essential to the understanding of the normal range, in addition to the interrelationships.

John Huntley, professor of English, says that knowledge is created when it is in a form that can be shared with others. That form may be a piece of text. Increasingly it will be in an interactive, multimedia form, perhaps a database, perhaps a mediated report of the traditional experiment. Such creations will occur among both faculty and students.

Knowledge is first created in the act of constructing relationships, and those constructions will be in a computer based form which can be altered to reflect new insights as the students learn. Their knowledge will become more complete and more clear as the semesters proceed, and they will be able to reveal what they have learned by developing multimedia term papers.

Finally, many students graduate to become teachers themselves—in our public schools, community colleges, and other colleges and universities. Those who are students today and teachers tomorrow will be more likely to integrate multimedia into their own curriculum if they are provided access to it while they are at the university.

## Computer Graphics: a Tool to Gain Insight

In the research environment, information is gained through models, simulations, and experiments. These all involve extensive calculations and produce massive amounts of data. The use of high quality computer graphics tools and techniques to visualize this data is essential for discoveries and increased insight. Moreover, the research component in a university is an important facet of the educational environment and cannot be separated from "education." Research is dependent upon student assistants, both undergraduate and graduate, many of whose theses will grow out of the research. A professor's research becomes student assistants' learning experiences, and the results from the research will find their way into national presentations and into the classroom. Color Plates 3–5, selected from an animation sequence, illustrate research on fluid mixing by Dr. Christoph Beckermann, University of Iowa. The color palette represents the mixture of clear water and salt water in a tank as heat is applied to the left side. Visualization was done on a Macintosh IIcx computer using NCSA Image software.

Research labs push all the limits of current computer technology. They demand high resolution, thousands of colors, and real time interaction with simulations or experiments running on remote machines. These require faster networks, faster computers, better color and resolution, and better user interfaces for interactions. Research is generally the first area to get high quality technical equipment because it gets more outside funding. However, issues of increasing computer speed and resolution, as well as network speed, and representing some forms of data, such as multivariate data, are research topics themselves.

What do all these research concerns have to do with computer graphics for education? Besides the fact that research labs provide the educational environment for student assistants, the classroom needs for computer graphics will also push one or more of these technological limits at some time. Student labs and classrooms need the same capabilities as research labs.

## *Need for Interaction*

Students can learn better through experience than by hearing about a topic or by passive observation. They will gain new insights by interacting with simulations and making their own discoveries. While discoveries gained this way may not be new to the world, they are new to the student. Students must have adequate hardware and software so that they can steer these experiments, making frequent changes to see what results. Design students can work with the parts of an object, make many changes, and immediately see the effects of these changes. Mathematics students can manipulate a formula and immediately see the graphical representation which results from the new formula. Science students can experiment with chemical combinations, different gravitational forces, or representations of solar systems. They may even learn through simulated disaster caused by an incorrect choice. Because there are more ways to learn than we now understand, we should expect to see more use of multimedia and increased forays into new areas of virtual reality. Interaction is essential, and the appropriate user interface is important, lest the computer get in the way of the students' exploration and interfere with students' natural curiosity.

*NEED FOR HIGH RESOLUTION AND REALISTIC COLOR*
Many classroom applications require high resolution and realistic color. Such applications range from interactive, multimedia applications for art or medical students to molecular modeling applications for chemistry students. The integrity of the images is essential here. Subtleties in color and shading must be observable. In most cases the interaction is still important, and high-quality images are essential. In a few cases, the interaction will have to be sacrificed for the highest quality images. In other cases, the quality of the images may need to be lower in order to get the best interaction.

*NEED FOR FASTER NETWORKS*
In many cases, the informational database with which the student is working is so large that it must reside on a central server. Examples of such large databases are geographic and cartographic data. Students interact with this data through a campus network. With current technology and large remote databases, students must wait a few minutes after a search, while the new data or images are being downloaded, so they can redraw the geographical representations. Faster flow of data would allow more inquiries and more exploration, and therefore result in more learning. Other needs for high speed networks include software servers, shared hardcopy and video devices, and simulations involving two or more students. It is possible, for example, for students from two different universities to collaborate in the creation of a video by sharing image files across the Internet, or to interactively and simultaneously collaborate on an art piece.

## Computer Graphics: a Tool to Provide Information Through Presentation Materials

There are several techniques for displaying information. These include color, three-dimensional modeling, and animation. A three-dimensional model allows viewing from different angles and potential virtual environment capabilities of walking through or flying over the object or structure.

Sometimes realistic, photographic quality color is important. More frequently, pseudo-coloring is employed to provide information about height, temperature, density, or any given variable. Depicting additional variables at a given time requires a combination of color with other capabilities, such as a three-dimensional model, stereoscopic view, sound, or even touch.

Some information cannot be detected with a static image, or even a series of images, unless the images are animated. Dan Sandin, Co-Director of the Electronic Visualization Laboratory at University of Illinois, Chicago, provides an excellent example of this phenomenon [4], and it is also evident in the random cellular automata research by Kostos Georgakokas at the University of Iowa.

In order to provide information, hard copy such as color plots, slides, and video from computer images is essential. These give researchers, teachers, and students media by which they can share their insights. This might be at the level of *peer* graphics, where the quality need only be high enough for a discussion among colleagues, or *presentation* graphics, where only extremely high-quality, polished video or images will suffice. Slides and videos can also help to explain experimental results to funding agencies, which is essential in order to obtain further funding.

For classroom use, a variety of presentation materials is also necessary, due to the variety in subject matter and complexity of images. It will become increasingly valuable for faculty members to be able to create their own educational videos for classroom use. For example, Joe Kearney's computer science robotics class visualizes constraint based robotic movement. Advanced rendering capabilities, including transparency and shadows, are essential to understanding the movement, as illustrated in Color Plate 6, a simulated flip.

The simulated flip in Color Plate 6 presents selected frames from an animation by Kearney and Hansen in Computer Science. The motion was generated through the mechanical simulation system, *Newton*, which automatically formulates and iteratively solves the dynamic equations of motion from a description of the hopper model. The hopping motion is controlled by constraints defined on the external forces and torques applied to the hopper as it pushes against surfaces in contact. By focusing on the interaction between the hopper and surrounding surfaces, Kearney and Hansen are able to derive device independent control programs that can be applied to a variety of hopper designs.

Kearney and Hansen also need the ability to create videotapes of their animations for class use. Color Plate 7 is a frame from an animation of a brigade of eight marching robots moving in groups of four. The motion was composed by duplicating and transforming a single-motion sequence generated through mechanical simulation. The animation demonstrates how a complex animation can be created by editing motion sequences. They have developed a system based on data flow programming in which a collection of independent motions can be edited, enhanced, and integrated to create and visualize dynamic worlds. This motion sequence was created by Dinesh Pai, and visualization of both robot images was done by John Knaack using Alias software on an IBM RS/6000 Workstation.

As the prices for a workstation and video output capability drop, and software becomes less obstreperous to use, these capabilities become more and more feasible. Faculty can produce VHS or S-VHS video with appropriate titling and sound themselves, perhaps at a central facility, perhaps in their own offices. Macintosh IIs are now affordable for office use, and powerful graphics workstations such as those from Hewlett-Packard, Sun, Silicon Graphics, and IBM are now affordable for departmental research labs. A chemist with such a workstation can create videotapes from his or her molecular modeling and analysis experiments for classroom use. In this case, the quality of the videotape must be as high as possible because of the complexity of the images and the subtleties in the shading. This professor would be likely to send the images over the campus network to a departmental or central site to be put onto tape, rather than creating his or her own videotapes.

# Computer Graphics: a Medium for Creative and Artistic Expression

Computer graphics allows students the capability to create and alter images and to generate prints, slides, and videos from those images. In addition to the use of computer graphics to obtain or transmit information, computer graphics serves as a unique medium for artistic expression. In some cases, art students will use computer graphics as a design tool to create the designs which will then be implemented through more traditional forms such as sculptures or prints. In other cases, they will take advantage of computer graphics as an artistic medium which allows capabilities different from the traditional tools such as pencils and paint brushes. This topic is elaborated upon by the artists in other chapters of this book.

## *Need for Video*

In each of the areas where computer graphics is needed (to gain information, to provide information, as a means of artistic expression), we see the need for computer animation and the ability to record these animations onto videotape. The needs for researchers and for professors have been mentioned. In some areas of scientific studies, for example, animations can provide much more useful information than still images and can show higher dimensions more fully. Students also need this capability in many areas. For example:

- Art students need to work with both two-dimensional and three-dimensional graphics, sometimes need to work with high resolution, realistic systems, and need to get high quality video animations of their work.

- Television departments train students in the making of videos for television. These students need to learn how to manipulate two-dimensional and three-dimensional graphics and need to have experience with current technology.

- Education students will create and use all kinds of audio-visual equipment in their careers. They need to learn how to produce interactive lessons, how to use existing technology such as videodiscs and Hyper-Card applications, and how to produce slides and videos for classroom use.

- One frequently appearing piece in the university mosaic is the departmental computer graphics class. This may be in art, computer science, geography, engineering, business, or education. These classes all have different goals and structures, but they all produce computer generated images. Students must be able to produce decent hard copy (paper, slides and video) of their images. These images may be part of class work, part of their résumés or portfolios upon graduation, or submissions to the annual SIGGRAPH Student Poster and Animation Contest and Exhibition (SPACE).

## *Summary*

Universities are becoming increasingly dependent upon the new information gained through the use of computer graphics and the capability to transmit this information to students through multimedia, interactive educational software, and high-quality slides and videos. These are discussed more fully in [1], [2], and [3]. Artists also find the high-quality computer graphics tools to be important for creative expression. White chalk on the blackboard is no longer adequate in higher education. Today's blackboard is multi-faceted. The myriad of facets encompasses all disciplines, the range of graduate and undergraduate studies, and the closely related classroom and research needs.

Capabilities that were available only through central facilities are now found in departmental research labs. These same capabilities are essential for the classroom in the near future. Along with the high-quality computer graphics workstations, the need for the ability to produce interactive multimedia applications and high quality videotapes across a high-speed network cuts across disciplines and academic levels.

## *References*

[1] Brown, Judith R. and Steve Cunningham, "Visualization in Higher Education," *Academic Computing*, March, 1990.
[2] Cunningham, Steve, Judith R. Brown, and Mike McGrath, "Visualization in Science and Engineering Education," *IEEE Tutorial: Visualization in Scientific Computing*, Gregory M. Nielson and Bruce Shriver, eds. IEEE Press, 1990.
[3] Zimmermann, Walter and Steve Cunningham, eds., *Visualization in Teaching and Learning Mathematics*, Mathematical Association of America *Notes*, Number 19, 1991.
[4] Sandin, Daniel J., "Random Dot Motion," ACM SIGGRAPH Video Review, Issue 42, July 1989.
[5] Contact for further information on Second Look Computing: Joan Sustik Huntley, Weeg Computing Center, University of Iowa, Iowa City, Iowa 52242. jhuntley@memex.weeg.uiowa.edu

# Remarks on Mathematical Courseware

### Charlie Gunn

*"What is now proved, was once only imagined."*
WILLIAM BLAKE[1]

Computers are increasingly present in mathematical research and education. In research, they have opened up new territories previously inaccessible and spawned new fields; in education, they are being hailed as a revolutionary teaching aid. These two trends have, for the most part, developed separately. There is, however, a common thread which can bring them together: each depends for its success on discovering ways in which computers can support mathematical *intuition*. We sketch the role of intuition in the process of doing mathematics by contrasting it to logic, its polar principle. Looking more closely, it is possible to distinguish *four freedoms* which, when respected, allow a software tool to support mathematical intuition, whether its use be research or education. These are the freedoms.

- to roam,
- to link,
- to vary, and
- to view.

Existing trends in software and hardware technology can be understood in light of their relation to these freedoms. Key trends which are discussed include:

- hypertext.
- object-oriented programming,
- interface design and construction, and
- three-dimensional graphics.

We suggest a rough correlation of these technologies with the four freedoms, in the respective order given, and indicate how courseware for geometry might be developed in the near future.

# Research and Education in American Mathematics

There is a growing recognition that the health of mathematics in this country is endangered by a failure of the educational process to attract the best and the brightest to the study of the subject.[2] One of the contributing causes is that there exist walls separating mathematicians who do research from those who do education. I claim that at least in one area, the realm of computers and mathematics, these walls should fall. This statement is based on the observation that the successful use of computers in mathematics depends on discovering ways to enlist the logical power of the computer in the service of the uniquely human capacity for mathematical intuition. Such a goal is best served by a unified program serving both the research and educational communities. This program will also yield benefits in the natural sciences, since they are built on mathematical foundations. I begin by a brief discussion of the roles played in mathematical activity by intuition and logic.

## *Logic and Intuition in Mathematical Activity*

There is not space to give a detailed discussion of the respective roles of logic, on the one hand, and intuition, on the other, in mathematical activity. This has been done elsewhere.[3] I will limit myself to a single quote from Henri Poincaré which sums up the position adopted here:

> You have probably seen the delicately assembled needles of silica which form the skeleton of certain sponges. When organic matter has disappeared, there remains only a frail and delicate lace. True, there is nothing but silica, but interesting is the form that it takes: we cannot understand it, if we don't know the living sponge which itself imprinted that form onto the silica. In this manner the old intuitive notions of our fathers, even if we have abandoned them, still imprint their form on the logical scaffolding we have put in their stead.
>
> This global view is necessary for the inventor; it is also necessary for he who wants to truly understand the inventor. Can it be obtained by logic? No, it cannot. The name which mathematicians give to logic will suffice to prove this: In Mathematics, logic is called Analysis, and analysis means division, dissection. The only tools it can have are thus the scalpel and the microscope.
>
> Thus, logic and intuition both have their necessary role. Both are essential. Logic, which alone can give certainty, is the instrument of proof; intuition is the instrument of invention.[4]

Here, "invention" can refer either to the act of original discovery or to rediscovery by a colleague or student. In either case an understanding of the experience involves more than logic.

How can we characterize intuition? We are faced with a common prejudice bred by an overly logical education, that intuition is somehow less real or exact than logic. To many people, the word connotes something out of the ordinary, unpredictable, inaccurate, phantastical, or simply wrong. To counter this, we can only bring forth the testimony of eminent mathematicians, such as Poincaré, who consistently have acknowledged the fundamental role that intuition plays in their discoveries.[5]

We can form an image of intuition by listing some of its qualities; it

- finds patterns;
- makes new connections;
- synthesizes and integrates;
- is playful and exploratory;
- is Gestalt-oriented ;
- works with large amounts of unfiltered information;
- happens in the human not in the machine

What sorts of qualities would we expect to find in a computer tool which aims to assist the human intuition in mathematical explorations? From the qualities listed above, we see that the intuition aims to overcome the existing logical framework of mathematics by making new, previously unsuspected connections. To this end, the human desires from the machine the widest possible freedom to explore and rearrange the existing map of mathematics. This freedom will allow playful exploration and construction of new connections between previously uncorrelated phenomena. The more that the computer is permeated with freedom, the more the human investigator will be able to exercise his intuitive powers.

## The Central Challenge

The fact that logic alone cannot circumscribe the world of mathematics was recognized theoretically in Gödel's Incompleteness Theorem at about the same time that the modern computer was being created. Based on the preceding discussion, we can go beyond this purely negative conclusion to suggest that the logical capacities of the machine can be enlisted in the service of human intuition. If this were not so, then there would be very little of interest in the task of transporting mathematics into the computer, since the logical structure of one is is mirrored by the logical structure of the other. However, once we propose that the finished product support intuition, we have made the task once more very challenging. This discussion can only begin to touch on the many and profound issues which arise.

## The Four Freedoms

Freedom, then, is a key ingredient of successful intuition. Studying the phenomena more closely, we can distinguish a set of four freedoms, which are associated with successful intuition.

### THE FREEDOM TO ROAM

Intuition wants to be able to get from one place to another without carrying a visa, climbing a wall, changing its clothes, or other unnecessary delays. Hence, the freedom to roam means an absence of all obstacles to navigation in the conceptual space of mathematics. It also implies there are multiple paths from one point to another. The student of Fourier analysis should be able to skip over to a biography of Fourier, shift from real Fourier analysis to the complex case, and, if that leads to group representations, be able to dive in there too. The freedom to roam also includes the freedom to jump, wander, meander, run, fly, survey, and — perhaps most importantly — to return.

### THE FREEDOM TO LINK

The intuition wants to be able to take "foo" and "bar" and create "foobar." The computer tool should support Gauss when he takes "sqrt" and −1 and comes up with sqrt (−1). Making connections should be possible even under difficult circumstances. This goal can probably never be realized, since there will always be new connections or inventions that supersede the tools themselves, but "man's reach should exceed his grasp, or what's a Heaven for?"[6] There are many different kinds of links, or connections, which the intuition makes. Not all are as dramatic as the invention of imaginary numbers; simply jotting something in the margin of a book is another sort of link. The freedom to link includes a spectrum of linkages.

*The Freedom to Vary*
Another way of making a connection between two states is to vary continuously from one to the other. In the language of science, this is parametrization. It means being able to put a knob or knobs on a configuration and then turn the knobs. This is something that computers do wonderfully. Unlike humans, they are quite happy to repeat the same experiment thousands of times with only slight changes. The human, on the other hand, excels at looking at the results and seeing patterns. It is largely the success of this process in the past twenty years that has led to the birth of a new field, tentatively known as experimental mathematics, in which the mathematician, using the computer as his lab instrument, acts like an empirical scientist, collecting evidence and forming hypotheses which he then attempts to prove.

One of the revolutionary achievements of the nineteenth century mathematics was the discovery of the projective equivalence of conics. Computer tools ought to allow us to rediscover such a parametrization naturally and easily.

*The Freedom to View*
Strictly speaking, no one has ever seen a number. Before we can see a number, we have to choose how to represent it. The freedom to view is concerned with the choices involved in this representation. Strictly speaking, it doesn't have to be a visual representation; a sequence of numbers can appear as ordinary text, as a graph of a function, or as a sequence of musical tones. It doesn't have to be a direct representation either; the same function could be fed through a filter before viewing or represented as a single point in a view of a function space. In general, any mathematical object that we are manipulating comes with a choice of bodies, and we get to choose which appears in a given experiment. We even get to design new ones.

## *Freedom Today*

These freedoms, as with other sorts of freedoms, may never be totally realized. But we can use them as a yardstick of progress and as compass points toward which we can orient current work. Where do we stand today with respect to these freedoms? We begin by noting that although better hardware is crucial to the process under discussion, we will ignore it; we assume that the machine can talk to all other computer hardware naturally and quickly, and it has the power to compute and display its results in near real-time for the types of tasks under consideration. We are interested here in how computer software can support intuition. By software we include both research software and educational software, also known as courseware.

As noted above, exporting mathematics onto the computer in a useful way depends on how well these freedoms are preserved. Thus it is not enough to write specific programs to compute specific mathematical results. And it is much more than archiving all mathematical writings in a computer-readable form. These tasks are vitally important to the success of the venture but not sufficient in themselves. Attention to the intuitive mode of both researcher and student requires the invention of new modes of access to this knowledge. These new modes will take advantage of the unique capabilities of the electronic computer in combination with all the human senses, including intuition. Our success in this endeavor may revolutionize how we look upon the activity of *doing mathematics*, a transition already begun by the advent of hand-held calculators. It will at the same time, as a derivative follows the curve, revolutionize the activity of *teaching mathematics*.

## Logic in the Service of Intuition: Existing Software Tools

To begin with, anyone who argues for the shining future for computational tools in mathematics had better address the current tarnished image of computing practice. There is a general sense that software

- takes too long to write,
- takes too little time to throw away,
- runs on too few machines, and
- is too difficult to extend.

All these problems are endemic to the software industry and vigorous efforts are under way to solve them. We review four main ingredients to the evolving response to this challenge:

- hypertext,
- object-oriented programming,
- interface design and construction, and
- three-dimensional graphics.

We conclude by showing how they relate to the four freedoms.

## Hypertext: A New Road

How might computer tools for mathematics appear in the future? Certainly it is possible to imagine that all mathematical literature has been somehow read into a computer. This process can be thought of as modernizing an old road. This in itself will not provide any new support for intuition. Beside this old road, however, a new one can be constructed, one that, to begin with, visits exactly the same towns. Like a modern Encyclopedia of Mathematics, it is comprehensive, compact, and precise. But instead of being written on paper and transferred to the computer, this encyclopedia is written for the computer from the start. It has articles which can be read, just like the old encyclopedia, but it offers much more to the reader than static pages of words and diagrams. The diagrams it offers the reader are live diagrams which can be brought to life by a standard set of interface gestures. It can equally well be thought of as a set of computer routines, tools, and data structures which allow the user of the encyclopedia, at whatever skill level, to construct his own examples and experiments. The user, exercising the freedoms of roaming, linking, varying, and viewing, is able to undertake explorations which correspond to his unique interests and experience. He is able to build upon what is there and leave a trace of his path for others to follow. In this vision, it is not necessary to distinguish between the researcher and the student, since as intuitive explorers they are indistinguishable.

To use another analogy, the inherited, logical foundation of mathematics is the DNA of mathematics, distilled into its clearest form. Like DNA, this nucleus is surrounded by a whirling dynamism, the metabolism of learning, where the intuition is called into play. At this level, the abstract content appears in a changing panorama of concrete, visualizable form. This level, the user interface, is infinitely plastic.

This vision of the future is based on a current software technology known as hypertext. As more and more knowledge of all kinds is accumulated in the computer, new ways of organization and navigation of this knowledge becomes more and more important. Otherwise the knowledge lies inaccessible and fallow. Hypertext is a vague but exciting concept that has arisen in response to this challenge. It aims to go beyond ordinary text, providing a variety of interconnections only available in an electronic system. Some of these facilities include:

- inclusion of live diagrams in documents;
- links between otherwise remote units of knowledge;
- layering of documents, to allow different levels of detail; and

- navigational aids for accessing large and complex databases.

These tools will support the creation of a mathematical encyclopedia which will take full advantage of the logical power of the computer, while supporting the intuitive leaps of the investigator.

## Object-Oriented Programming Style

As an introduction to this topic, let's begin with a question: What is the limit of how much of mathematics can be imported into the computer? Today, anyone who wants to do mathematics on the computer has to build on top of very few primitives — the integers, the reals, and, in some implementations, the complex numbers. This is the current limit. Manipulating a more complicated object, such as a group or a space curve, requires a large programming effort, and this effort is not easily transferred to others who want to do similar computations.

This is the case, in large part, because we don't have a uniform format for these objects that is portable between programs, machines, and operating systems. Who would want to use a computer if he had to personally write the routines for manipulating floating-point numbers? Designing and implementing formats, or languages, for mathematical structures is a daunting task. We need only think of how many hundreds of man-years went into the refinement of the floating point standards which are now accepted across the industry. But it will eventually enable a mathematical programmer to access an abstract group as a high-level computational object with the same ease with which he now manipulates ordinary numbers. These computational objects will know how to perform appropriate operations on themselves, how to write themselves to files and how to retrieve themselves. This method of structuring programs is known as object-oriented programming. For example, a researcher will be able to make one of the 230 three-dimensional Euclidean crystallographic groups act on a geometric object and display the result with no more effort than it takes now to perform an ordinary multiplication and print the result. Researchers working with dynamical systems, which includes many scientists in other disciplines, will be able to communicate their experiments and results using very compact expressions which can be understood by other computers, just as today all computers understand floating-point number formats.

The key term here is *object-oriented programming*. As indicated above, this type of programming language supports hierarchy and modularity through features of the language. This is particularly important for those who are faced with translating mathematics into the computer, since mathematics is heavily hierarchical. In the same way that the current system of journals provides a sturdy vehicle for the evolution of mathematical heritage, these shared computational objects would become part of the practicing mathematician's repertoire, which he would build upon and refine.

## *Application Interface Design and Construction*

One of the most important ingredients of computers in the service of intuition is simplifying the task of software development. For example, practicing mathematicians want to be able to develop their programs in a matter of hours, not days. A first step in this direction is the specification of standard programming environments. What does such an environment consist of? Most include a windowing system, whereby input and output to the computer occurs via graphics-rich and event-driven windows. Windows include not only conventional windows for typing text but also a variety of iconographic regions such as sliders, dials, buttons, and menus which allow non-textual input and output. One level above this there is now a Graphical User Interface (GUI), which specifies the appearance and behavior of these windows. A GUI allows easy manufacture of standard interfaces across a wide spectrum of environments. The next few years should see a maturation in this field and the emergence of usable, standard environment for window-based programming.

Specification of the GUI, is, however, only the first step to strengthening intuition. The next step is to make it easy to create applications which are elegant, easy to use, and visually interesting.[7] Once we acknowledge that we want to aid the intuition, then the ease-of-use and interactivity of a program become legitimate criteria for its success. The care and craft that go into the creation of programs to aid intuition deserve the same attention and recognition that the articulation of theorems and proofs receives. Hence, it is important to be able to create such programs as quickly as possible. Graphical interface and object-oriented programming techniques are being combined to provide such tools, known as interface builders, for automatically generating the user interface to programs. This paradigm gives the programmer a menu of commonly used interface objects, such as menus, sliders, buttons, dialog boxes, and animation controls. From this menu the programmer can design the interface to his program, much as a young person playing with a punch-out book of clothes can dress the cardboard doll as he chooses. The interface builder supplies a skeletal source code-program into which the programmer need insert only his application-specific code.

## *Three-Dimensional Graphics*

Much has been written about visual thinking and the connection of visual perception with mathematical concepts.[8] This discussion has been stimulated by the tremendous growth of computer graphics technology which makes possible the display and manipulation of complex three-dimensional objects. In a few years, even inexpensive computers will have this capability. One of the great drawbacks, however, is the absence of any standards here, so that graphics programs tend to be non-portable. There has been lively activity in standards in recent years, and the advent of standardized three-dimensional scene descriptions is inevitable. Such a standard will unleash a powerful new force into the toolkit of the mathematician. It will serve for three dimensions the same purpose that PostScript now serves for two dimensions: as a device-independent, nearly universal description language. One of the challenges for any such standard will be how much freedom it allows the mathematician to visualize the previously unseen realms of mathematics. Every practicing mathematician can contribute one or more mathematical topics from his own experience as teacher or student which could be better understood through three dimensional graphics.

## *The Four Ingredients and the Four Freedoms*

It is interesting to correlate these four ingredients of today's software environment with the four freedoms introduced before. To a certain extent there is an affinity between pairs. For example, hypertext is primarily designed to support the freedom to roam. Graphics technology is the primary expression now of the freedom to view. Object-oriented programming is especially helpful in getting disparate objects to relate with each other, which is perhaps the most important example of the freedom to link. Finally, interface builders are primarily used to design control panels. These, in turn, are used to vary the parameters of an experiment. The matching isn't exact but provides a useful way of thinking about the subject.

Table 1 encapsulates these correlations.

| | | |
|---|---|---|
| To roam | 🗘 | Hypertext |
| To link | ↗ | OOP |
| To vary | ⊘ | Interface |
| To view | 👁 | Graphics |

**Table 1: The four freedoms and the four technologies**

## *Consequences for Courseware*

These developments will also bring powerful benefits to the field of mathematics education. Imagine that the fundamental structures of mathematics are incorporated into the computer in an object-oriented way and are connected to sophisticated user interfaces and three dimensional graphical representations. It will then be possible to write robust, powerful courseware in substantially less time than it takes now — the time will be spent on the contents of the course, not on programming it. We reiterate: once a mathematical structure has been implemented in a computer, it is increasingly easy to choose how to clothe it and how to present it to the user. This represents another argument for researchers and educators to unify their efforts to put mathematics on the computer.

Current work with the Geometry Supercomputer Project has provided some specific themes. A first step has been the creation of the Object-Oriented Graphics Library (OOGL). The objects it knows about are three-dimensional surface representations, and the operations it knows how to perform are primarily graphical ones. Within this limited sphere, it provides an easy way to create and operate on many types of geometry, including polygons, meshes of quadrilaterals, parametric splines, and vector lists. On this base a general purpose 3-dimensional viewing tool has been constructed. This allows the mathematician to write interactive applications which deal only with geometry, not graphics, and invoke the viewer as a scientist would use a microscope to examine his experiment. Using this library, the project has created a variety of applications:

- a simulation of a spinning top,
- investigation of Dirichlet domains of hyperbolic and Euclidean 3-manifolds,
- a surface evolver for minimal surfaces, and
- a link evolver to find good embeddings of knots and links in 3-space.

In the future, the plan is to expand the set of objects supported by OOGL to include more mathematically interesting ones, include simplicial complexes and 3-manifolds. OOGL will become more the Object Oriented *Geometry Library*. Courseware applications might begin by implementing Klein's Erlanger Program[9] in an object-oriented way, providing software to study the spectrum of geometry from projective to affine into metric, including Euclidean, elliptic, and hyperbolic.

## Conclusion

We have shown how the challenge of translating mathematics into the computer can be understood as a task in enlisting logic in the service of intuition. From this point of view, both research and education are equally well served by its successful completion. All the sciences, too, will benefit to the extent that they all built upon mathematics. The strategy outlined above is to develop software tools which support the four freedoms of roaming, linking, varying, and viewing. Embryonic forms of these tools are available now as hypertext, object-oriented programming languages, interface builders, and three dimensional graphics. The mathematics community, both researchers and educators, can take advantage of these new developments by beginning to work together to incorporate them into courseware for both research and education.

## Notes

[1] One of the *"Proverbs from Hell"* in *Marriage of Heaven and Hell*, 1793.
[2] See, for example, David, Edward E. Jr., "Renewing U.S. Mathematics: An Agenda to Begin the Second Century," *Notices of the AMS*, 35 (October 1988), 1110–1123.
[3] See, for example, *The Creative Process*, Brewster Ghiselin, ed., The New American Library, New York, 1952.
[4] Henri Poincaré, *La Valeur de la Science*, quoted in Detlef Hardorp, "Why Learn Mathematics?," 1982, preprint.
[5] Ghiselin, *The Creative Process*.
[6] Robert Browning.
[7] Other senses not excluded.

[8] See, for example, *Visualization in Teaching and Learning Mathematics*, Walter Zimmermann and Steve Cunningham, eds., *MAA Notes* 19, Mathematical Association of America, Washington, D.C., 1991.
[9] For a description of this, see Carl Boyer, *A History of Mathematics*, Princeton, 1968, p. 593.

# Visual Ways of Knowing, Thinking, and Interacting

### Kenneth O'Connell

People can think and work visually, and computer graphics hardware and software are making important contributions to the visualization process. Visual thinking is a part of our basic perceptual and mental processes. Visual ways of knowing and working are important in our educational environment. They are partners to the traditional verbal and mathematical ways of knowing that we teach and encourage. The software that is being developed today and in the future must be very creative, interactive, and work with integrating multiple ways of knowing. It must be visual and use sound, sequence, and animation to communicate ideas. We must be able to manipulate and play with the form and content of our explorations and learning.

## Visual Thinking by Artists and Scientists

We have a great deal to learn from the visual thinking methods of artists and scientists as we explore the increasing role that computer graphics can play in our educational process. In studying visualization and visual thinking, we find that history shows us many examples of scientists and inventors who were also artists. These people profited by using their education in the arts to increase their observational abilities and fine-tune their ability to analyze. Samuel Morse, the inventor of the telegraph, was educated as a painter. Louis Pasteur also studied painting. Robert Root-Bernstein, of the Center for Integrative Studies at Michigan State University, states that there are shared ways of thinking used by both artists and scientists [7].

Root-Bernstein has done extensive research into the lives of important historical figures who developed talents in both the sciences and arts. He refers to twelve tools of thought that help us use our imagination to observe, analyze, and explore.

1. *Pattern Seeking*. The artist as well as the scientist places great value in seeing patterns and hidden structure where others do not see any. We search for patterns to discover principles and reveal information.
2. *Pattern Forming*. By creating patterns and by bringing ideas and images together we see our thinking more clearly. Scientific visualization is based on forming patterns with the data to be able to see new relationships and new patterns.
3. *Analogizing*. The ability to see how an idea in one domain or subject area can be related or transferred to another domain or subject can help us understand the concept or a new relationship. Thinking of DNA as "unzipping" helps related common objects to invisible structures.
4. *Abstracting*. To abstract is to eliminate all the unnecessary details from our perception and reveal the basic principle or structure.
5. *Mental Visualization*. Seeing with the mind's eye is our most direct way to create images of what we are thinking about. Trying to picture something helps us to work out its structure. Some people are able to animate their visualizations and see how they move.
6. *Physical Modeling*. Our ideas need a three dimensional form because it is so hard to think in three dimensions. Sculptors do it better than anyone else. Architects and molecular biologists must design with space in mind for new buildings or new molecules.
7. *Kinesthetic thinking*. Some people simply need to move when they think or are learning. Many learning skills are imbedded in muscle memory. Driving a car, using hand tools, throwing pots on a wheel, and computer interactivity build on the motion memory that is being built up in our bodies.
8. *Aesthetics*. We are often caught off guard by the sudden recognition of the beauty of an image, view, or experience. Beauty has a wholeness and an emotional feel that commands all of our attention. What do we see when we are moved by beauty?
9. *Playacting or Internalizing* (imagining oneself as an object). Identifying with an object or concept helps us to know it better and begin to imagine interacting with it.
10. *Manipulative Skill* (hand-eye coordination). By working with materials, by sketching, and by building models we derive knowledge through our senses.

11. *Playing* (experimenting: trial and error). Sometimes, as a way of learning, we must give ourselves permission to try something at which we may fail. Play and humor help us feel more comfortable with what we don't know so we can explore what we may find.

12. *Transformational Thinking*. Change is a part of us and we transform as we grow and so does the world. The dynamics and forms of change are important to understand. What is the sequence? Where does it begin? What are the critical points along it?

Root-Bernstein concludes by saying that these tools are not useful unless the ideas can be communicated by yet another set of skills:

(a) verbal (written) communication,

(b) mathematical communication,

(c) visual communication,

(d) aural (sound) communication, and

(e) kinesthetic-tactile communication.

These methods of communication are for both receiving and expressing information. How many of these skills are ever considered in the development of computer software?

One of the ways that I ask people to experience first hand the use of visual thinking, pattern recognition, and pattern forming is to draw from memory. I ask students or members of the audience to make a quick sketch of a simple bicycle. Everyone has seen a bicycle and so this seems quite simple, but soon people realize that it is not so simple. They first draw the two wheels and their centers without much trouble. They stop as they draw the connecting bars and try to visualize the structure between the wheels. I ask them to speculate if they can't remember the details. They try to form a pattern of connecting these two nicely drawn wheels. After a few minutes I bring a bicycle into the room or show a slide of a bicycle (Figure 1), and everyone looks at it with a heightened perception and careful attention to the connecting bars. I ask them to draw it again, now that it is in front of them. I explain the basic diamond shape of the frame and discuss the problem of knowing a structure or pattern well enough to reproduce it compared to being able to recognize a general object without a structural understanding. Once they draw the bicycle and focus on the principle structure, it is clear and they will always be able to draw a bicycle. The visual structure or pattern is now known and can be formed or visualized without difficulty.

**Figure 1: The traffic sign shows a simplified yet quite accurate drawing of a bicycle to compare to the memory drawings**

## Visual Maps in the Mind

In a recent article John Allman suggests that "there is this whole bunch of visual maps sitting in the back of the brain." When we look at something or try to reconstruct an image we employ some of these maps. Each seems to have a different function. One map may respond strongly to shape, another to color, another to motion, and another to edges. He concludes that these mental maps contain learning and memory, rather than being separate from them. Perhaps the viewer could only recall the general pattern of the bicycle as two round wheels and some bars in between, so only the general pattern recognition map was working. After I taught people to draw the bicycle they had no trouble remembering it, as though the map for detailed line structure was now activated, linked with the old general shape map, and ready to fire when the mind needed it.

## Creative Software for the Imagination

A piece of software written by Craig Hickman for his three year old son, Ben, displays some of the most creative thinking in programming in a long time.

# Visual Ways of Knowing, Thinking, and Interacting

Craig, an artist who programs, started to write a simple paint program that would be frustration-free for his son to use. When he let his son try conventional paint programs, Ben would get lost or would accidentally click on a dialogue box that he was unable to read. So Craig wrote the program so that whatever Ben tried he could not get lost. The program evolved and now has full color and sound, and is commercially marketed as KID PIX™ by Brøderbund Software, Inc. My focus here is on the unique and creative features of this program and the implications for the future educational software.

The basic design of the KID PIX working window has the various tools along the left hand side with the options of the tool along the bottom. The overall KID PIX screen (with a paintbrush image) is shown in Color Plate 8 and a forest scene created with KID PIX is shown in Color Plate 9.

There are many creative aspects of this unusual paint program. First it is different from all others because it uses sound with every tool. You can hear yourself paint. Craig then set about to give each tool some personality or character, or as he referred to it, "I tried to give the tools some limited intelligence." Examples are the wacky brush that blobs like a leaky ink pen, the crooked pen that will not allow you to draw a straight line, and another brush has large drops that drip from each line (each splash comes with a sound to match it). A set of stamps and a set of letters will appear at the bottom of the screen to provide choice in building a picture. Each stamp has a sound, and each letter says itself in English or Spanish.

While most programs have simple erasers, this program has creative erasers that remove your image in fascinating ways, as shown in Color Plate 10. One eraser blows up your image with explosive sounds while another, called the "black hole," sucks up the image. With another eraser, the image breaks up into many rectangles that fall off the screen. Each eraser can be stopped at any point if you like its effect on your image. In fact, changing the whole image is the motivation for the set of image processing tools.

The egg beater is the icon for the image processing tools. These acquaint the user with the notion of altering an image by doing things to it in part or in whole. These tools do not basically create any image by themselves, instead they alter and change the images that have been created, changing the whole image by enlarging, shrinking, breaking it up into squares, adding edges, converting blacks to white or colors to their complementary. A typical user will go from creating an image in a brush pallet, change it in the erase pallet, and finally modify it with the process tools. These tools expand, shrink, outline, and multiply the basic image.

## Playing and Interacting

One of the most interesting and unexpected result of this program was the way young children use it to stories and tell stories. Young Ben would fill the screen with many lines, rubber stamps of a suitcase, and frog stamps. When I asked him about it he said that it wasn't a suitcase, it was a car and inside was a frog who was driving all around. The final image is of little value here, it is the interaction and the role playing during the process of using the program that are of value. Figures 2 and 3 show some of the creative results of using this program. Craig also wrote a routine that uses the sound-recording capability of some of the Macintosh computers. Young children have recorded short poems for their pictures, and when the picture is opened the poem is read aloud.

**Figure 2: A young child draws and adds text to this wonderful concept of the "fish jumps to the best part of the story"**

We know that being able to manipulate things helps us exercise our thinking. In many Japanese elementary classes a student has a "math set." This often consists of a box with materials to help in learning math concepts. The set has tiles, clock, ruler, checkerboard, colored triangles, beads, and many other interesting objects. With these objects abstract concepts can be given solid form, allowing pattern forming and pattern seeking.

## Future Implications

Many creative aspects need to be considered in the future design of educational software. They should challenge us to think. Craig Hickman's program captures the imagination of the user and helps the user move

# Visual Ways of Knowing, Thinking, and Interacting 135

**Figure 3: Two small children draw their favorite animal and type their name on the screen**

through his or her own creative ideas more easily. By giving us animated, sound connected, intelligent, and imaginative tools Craig shows us a rich assortment of ways to create images and enjoy learning.

This creative combination of very interactive tools, animation, and sound could be a model for the development of programs that help people explore such subjects as crystal structure, architectural models, geometry, and molecular models. Each program could have basic building units, assorted variations on structure units, linking processes, special sounds, and hidden information that keeps popping up. The user can then create and explore at the same time.

Another series of programs could be based on insects, their houses, and their environments. You could build and modify animated bugs as well as build complex houses for the various insects. Sound and image processing could help develop the weather, environment, and animal noises. Again all sorts of information could be built into the discovery process.

An engineer who saw this software demonstrated remarked that he would now start to use sound in the design of his CAD programs as a feedback mechanism to help the user know what tool they just clicked.

I've taught adults with this program and they loved it. They found it to stimulate creative and imaginative thinking. It caused them to think sideways by bumping them into variations of ideas that they wouldn't normally find if they went straight about their business. In fact it is hard to go straight about your business with this program as you get engaged with all the choices that you have. Adults also found that if they got stuck as they were trying to develop an image, all they had to do was to remember one of the tools that was sort of like what they wanted to do. They then switch to this tool and see if it will help them move toward their idea. The era of stimulating, interactive, imaginative, and educational software is here. Let's start writing more software that does what this program has shown us is possible.

## References

[1] Gardner, H., *Frames of Mind: The Theory of Multiple Intelligences*. New York: Basic Books, 1983.
[2] McManus, N., "Kid Pix Vrooms Past Sculley at Expo," *MacWeek*, January 29, 1991.
[3] Meltzer, B., "Who Can Draw With a Macintosh?" *The Computing Teacher*, April, 1990.
[4] Montgomery, G., "How We Really See, The Multi-Screen Theater in Our Brain," *Discover*, May 1991.
[5] Piaget, J., *To Understand is to Invent: The Future of Education*. Harmondsworth, England: Penguin Books, 1973.
[6] Root-Bernstein, R. S., "Teaching Abstracting in Integrated Art and Science Curriculum," *Roeper Review* 13(1991), 85–90.
[7] Root-Bernstein, R. S., "Tools for Thought: Designing an Integrated Curriculum for Lifelong Learners," *Roeper Review* 10(1987), 17–21.
[8] Root-Bernstein, R. S., *Discovering*. Cambridge, MA: Harvard Press, 1989.
[9] Stigler, J. and Stevenson, H., "How Asian Teachers Polish Each Lesson to Perfection," *American Educator*, Washington D.C. Spring, 1991.

# Visualization of Concepts in Physics

## Hermann Haertel

Teaching science, especially physics, is loaded with mathematical formalism and abstractions and visualization and qualitative models are lacking. The defenders of this traditional teaching approach declare that the formal description of nature is the only truly scientific language which is free of contradictions and which should, therefore, be taught from the beginning. In their view, pictures or qualitative models always have drawbacks, are of limited value, and risk leading students to false concepts and misunderstandings.

The argument against this point of view, and the force that drives the demand for qualitative models and a more visualized and less formal approach, is based on the opinion that a large number of learners construct images and use qualitative models whenever dealing with abstract representations and mathematical derivations. If this is the case, it should be much more efficient to treat these qualitative models explicitly, to criticize and to reconstruct them, in order to optimize the learning process. This long-lasting debate between advocates of different teaching approaches takes on a new aspect as computer graphics and powerful work stations become available. Interactive animated graphics seems to have a didactical potential which ideally bridges the gap between qualitative models on one side and quantitative methods on the other.

## Didactical Potential of Computer Graphics

In this paper we give some examples, taken from work on electricity as a special topic in physics, that demonstrate the didactical potential of computer graphics. It should be possible to see some general ideas in these examples which can be applied more widely.

### Construction of Basic Concepts

In traditional teaching of electricity, the elementary charge in the form of a point charge or particle is introduced as a basic unit of all electrical phenomena. This is visualized and treated as an independent and fundamental particle of nature. The electric field, a much more abstract concept, is introduced much later and is often described as though it is produced by the point charges.

It could well be that this difference in the treatment of charge and field was influenced by the difference in effort necessary to represent these concepts. It is quite easy to draw particles in arbitrary arrangements, but the appropriate field can only be drawn for rather simple or highly symmetric setups or on the basis of extensive calculations. With computer graphics, this difference has disappeared, and it is now possible to visualize the electrical charge and field always together. This is a more fundamental *gestalt* than individual point charges and separated fields, because in nature both appear, always related as two sides of a whole. Charges cannot be detected without field forces, and every field line leads to some charge.[1]

A basic didactic question can be raised here related to the traditional approach of introducing basic objects and concepts in physics. It is common to present those fundamental objects of nature to the beginner in a highly idealised form (such as point mass, rigid body, single force, or equilibrium states without transient processes) and to demonstrate the truth of basic laws of physics under these idealised conditions. Once these laws have been understood, higher complexity is introduced by weakening these idealisations in order to cover more realistic problems. Such an approach is based on the assumption that simple and idealised objects are also simple to learn, and it is proper therefore to start with simple and even highly idealised situations.

There is no question that in general a learning process should develop from the simple to the more complex. But is the simplest also the easiest? Point charges, or in mechanics point masses, are so idealised — objects without any dimension — that no further properties (such as volume or structure) can be associated with them. When dealing with such a concept, is active learning — a reconstruction of knowledge by the learner — stimulated and supported to the necessary degree by these simplifications?

Computer graphics is a new tool that allows the representation of objects with higher complexity, such as the representation of charges and fields as an indivisible unit. In mechanics it allows the representation of elastic bodies instead of rigid ones or point masses. The general questions to be raised in respect to this tool would be: What is the correct level of idealisation to start with? How can active learning be stimulated to a maximum? What is the tradeoff by raising complexity and a possible learning barrier at the beginning? Careful evaluation studies will have to show how this idea can be introduced into practice, and whether it shows the expected positive results in relation to better learning and understanding.

## *Representation of Particles*

There is another problem with the usual representation of particles. In nature you either find neutral objects or separated charges with fields around them. You do not find an electron as a particle with a defined surface. The representation of electrons as particles is therefore a problematic approach because it can lead to an early fixation and to difficulties when other representations (such as solutions of wave equations) have to be treated. A medium like computer graphics, which allows a rather flexible choice of different representations of such fundamental concepts, could offer a valuable tool to lead students to a less static or more flexible view and, therefore, to a better and more complete understanding of basic principles in science (Figure 1).

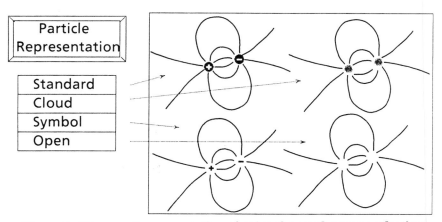

**Figure 1: Change of representation of point charges by menu selection**

## *Multi-Dimensional Representation*

The dominance of formal methods in traditional teaching of physics has another drawback. The students have to learn a new topic, such as the concept of field or the term "potential," and at the same time the use of a new language, such as vector algebra or differential equations. It could well be that the extra burden of a new topic together with a new language is at the

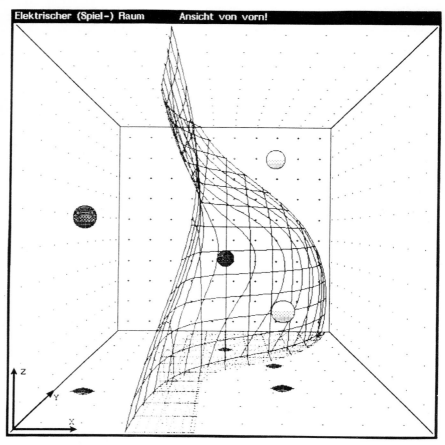

**Figure 2: Representation of charge carriers and equipotential lines in three dimensions**

root of many experiences of failure and, therefore, often the reason for losing interest in the field of science. The introduction of natural or technical processes in a three-dimensional interactive representation, which is now possible with powerful graphic workstations, eliminates this extra burden for a beginner (see Figures 2 and 3). It should make a big difference whether we see an abstract diagram or two-dimensional scheme and then reconstruct a three-dimensional picture in our minds, or go the other way around. With this latter approach, we see something in three dimensions which we can manipulate and from which we can construct representations of higher abstraction.

# Visualization of Concepts in Physics

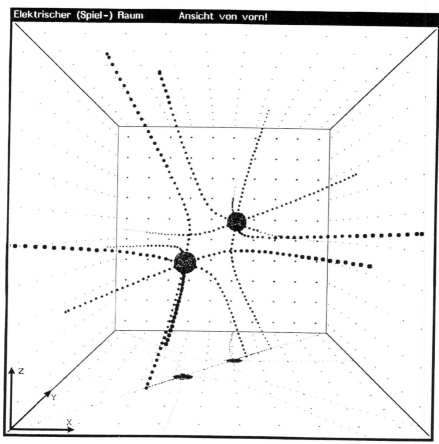

**Figure 3: Representation of charge carriers
and field lines in three dimensions**

In Figure 2, charge carriers and equipotential lines are shown as an example of this kind of representation. In addition to these, field lines can be drawn (see Figure 3), Coulomb forces can be indicated, and it will be possible to detect the flux, div, and curl in relation to any closed surface or sphere. It will be interesting to evaluate a course where electrostatic concepts have been introduced this way, compared to the traditional approach where diagrams and equations are used at the beginning.

## Linking Different Kinds of Representations and Descriptions

The traditional approach to teaching physics is heavily based on the demonstration of experiments, and there are good reasons for this. An experimental result is an indispensable criteria for verification or falsification of any statement or hypothesis, and, therefore, the experiment is indispensable in any serious teaching lesson. However, such an experiment is often burdened with too-high expectations in relation to its didactical value. An experiment rarely speaks for itself, and often you need to know what you have to learn in order to see it. Computer graphics in conjunction with an experiment can help to visualize the essential parts and highlight the important aspects and functions of the usually complex setup and can, therefore, strongly enhance the experiment's didactical value. In addition, today video technology allows the showing of a continuous display from an experimental setup to the underlying physics, which should further reduce the gap between different levels of abstraction and should, therefore, have a positive influence on successful learning and understanding.

## New Ways to Integrate Mathematics

Without the use of computers, elegant mathematical methods such as Fourier analysis and closed solutions of partial differential equations were the best way to come to a quantitative result in reasonable time. With the use of computers and adequate numerical methods, the number of necessary mathematical tools can now be drastically reduced while the range of application and the intelligibility for students is obviously increased. A simple example is the possibility of changing the exponent in the $1/r^2$ dependance of the Coulomb force and seeing an immediate display of the different results to find out the special importance of the existing law compared to the alternatives (Color Plate 11).

A more advanced example is related to high frequency phenomena, where all processes can now be treated in the time domain and no abstraction to the frequency domain (and Fourier analysis or Fourier integrals) is necessary for the beginner. This extra step can be left to advanced study where formal qualification has higher priority than causal explanation and deeper insight.

## The COLOS Project

The work described here is mostly being carried out within a European project named COLOS (Conceptual Learning of Science). Within this project, initiated by Hewlett-Packard in 1987, eleven university groups from eight European countries work together with a common hardware and software environment in order to develop teaching and learning programs for university-level studies. The RMG object-oriented programming environment, developed at Hewlett-Packard laboratories in Palo Alto and used within COLOS, includes the following tools:

- a text editor,
- an icon editor,
- a panel to design windows and subviews,
- an interactive class browser,
- an interactive message browser,
- an interactive instance browser,
- an interactive pointer browser,
- a UNIX subshell window, and
- message buttons for rapid prototyping on the screen.

While RMG is partly written in plain C and partly in Objective-C, it is possible to make use of the fast graphics of powerful workstations as well as of object-oriented programming and rapid prototyping. RMG thus narrows the gap between the major programming effort usually necessary to develop interactive user interfaces under UNIX, and the relatively minor effort given to solve the actual didactical problem. RMG, however, is still not a final solution to future programming needs. Most of all, it does not conform to standards such as the X Window System or Motif. However, it is planned to port RMG to the new HP-Apollo 700 hardware platform, integrated into standards such as the X Window System and other modern programming tools such as Softbench. The COLOS project, supported until now by the European COMETT program, will continue in a reorganized way and focus on the following activities:

- development of tools,
- development of modules for training, and
- evaluation studies.

The development of tools will include work on telecommunication and distance learning. The topics for the training modules will, at the beginning, concentrate on basic concepts in science, especially electricity. The development is open for other areas of interest when further support from the European Community can be found.

## Notes

[1] Electromagnetic waves seem to prove the existence of fields without charge (if one neglects the origin of such waves, sometimes light years away). In this case, however, no static components but only time derivatives of fields are present and the interpretation of those waves depend strongly on concepts about vacuum and empty space where further questions can be asked.

# Prospero: A System for Representing the Lazy Evaluation of Functions

## Jonathan P. Taylor

Prospero is a system designed to present the evaluation of functions written in a lazy functional language. Any system which displays the evaluation of a program written in any programming language must have

- a way of dividing an evaluation into a number of stages,
- a computational representation of a stage,
- a set of transformation rules to move from stage to stage,
- a way of displaying a stage to a user, and
- a way for the user to move through the stages of the computation.

I will describe the way in which all of the above features are incorporated into Prospero and will give a number of ways in which Prospero may be used as a tool to aid in the teaching of functional languages.

### Functional Programming

A functional program is a set of rewrite rules written in the form of equations. Given a set of equations, a functional programming system may be asked to calculate the value of an expression involving the equations. Lazy functional programming systems aim to carry out the minimal number of calculations in order to calculate the value of an expression. They do this in two ways. First, the calculation of any expression is delayed until its value is demanded. Second, a value that occurs more than once in a calculation is only calculated once and the result of the calculation is shared. A detailed explanation of functional programming can be found in Bird [2].

### Evaluation of Expressions

It is easy to see that the evaluation of a numerical expression can be represented as a number of sub-calculations leading to the final value. For example the steps in evaluating

```
        2 + (8 * (4 + 1))
could be
        2 + (8 * (4 + 1))
        =>2 + (8 * 5)
        =>2 + 40
        =>42
```

The calculation of any expression written in a functional language can be represented in such a way. When using a functional programming system such as Miranda[1], however, the programmer will never be shown these intermediate steps. The evaluation in the eyes of the user is, instead, one monolithic step. Every functional programmer must develop the skill of producing a set of intermediate steps of a calculation with pencil and paper.

*Pencil-and-Paper Problems*

During recent observation of novice programmers[2], I noted a number of problems arising when pencil-and-paper evaluation was adopted. These observations are provisional, and more observation is required before they can be stated categorically. The observations are discussed in more detail in Taylor [7] and can be summarised as follows:

- Misconceptions which introduce errors into function definitions are usually carried through as misconceptions in pencil-and-paper evaluation.
- Pencil-and-paper evaluation is a task which is too complicated to be carried out for anything but the simplest set of definitions.

Despite these problems, pencil-and-paper rewriting is a useful teaching technique; when it is used correctly by students it often leads to insight about the behaviour of their definitions. By representing the steps in the evaluation of expressions, I aim to help with the problems outlined above and reinforce the power of rewriting techniques.

## The Prospero System

Prospero can evaluate functional programs, has an interface which displays the stages of the evaluation to the user, has an abstract interface which is an internal representation of the interface, and has a navigation system which allows the user to move through evaluation histories. The overall structure of Prospero is shown in Figure 1. An evaluation history is represented by a sequence of reduction graphs. The user is able to have a number of concurrent views of the graph, each of which has a filter associated with it. With these filters, users may specify a subset of the information present in the graph; this subset is then displayed in the view.

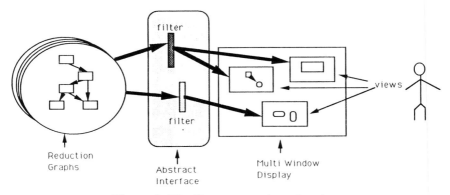

**Figure 1: The Prospero system structure**

## A Model for Evaluation of Functions

It has been suggested (Pain [4]) that when choosing an evaluation model for a language, it is essential to choose one which will faithfully represent the evaluation of all programs written in that language and which is close enough to the original definitions for the user to be able to understand which piece of a representation maps to which equation. The model chosen to represent the intermediate steps of the evaluation of a set of Miranda definitions is graph reduction. In graph reduction, an expression is represented as a graph. Reduction rules are then repeatedly applied to the graph until a graph of "canonical" form is reached.

Most implementations of functional languages are based upon graph reduction. Simple graph reduction is unacceptably slow and so implementations are heavily optimised; these implementations were not designed to provide debugging information, and so make no attempt to preserve any information removed by optimisations. It has been shown that such information can be retained with acceptable overheads (Toyn [9], Snyder [6]). But such research is not the goal of the Prospero project.

Instead Prospero adopts simple graph reduction. Prospero's set of reduction rules is a small one and can easily be expressed using example graphs and a simple English narrative in a way similar to the one described in Mobus [3].

GENERATING THE HISTORY OF AN EVALUATION
The history of an evaluation is generated in the following way:

1) Generate the reduction graph which represents the expression to be evaluated.

2) While the Reduction Graph is not in a canonical form

   a) apply the appropriate reduction rule to the graph to make a new graph,
   b) add the graph to the list of graphs generated so far.

Each iteration of step 2 is called an atomic reduction step. No graph in the history is generated until the user demands to be shown that part of the history, at which point all of the history up to the specified point must be generated. Navigation through the history is discussed in a later section. Details of Prospero's graph reduction model can be found in Taylor [8].

## *Displaying Reduction Graphs*

The use of reduction graphs as the sole representation of stages in an evaluation would be intolerable for two reasons. Firstly, the size of a reduction graph for anything but the most trivial expression is too large to allow the display of the graph in a meaningful way. Secondly, although the initial graph closely represents the expression to be evaluated, the size of the graphs produced quickly obscures their meaning.

The problem with reduction graphs is that they contain information at a very low level; that is, each node of the graph contains a very small amount of information. Presenting a reduction as a series of reduction steps is comparable, in some ways, to showing machine code to a PASCAL programmer.

FILTERS
To manage the potentially high levels of information present in a reduction graph, Prospero has a filtering facility. This facility allows a user to remove some of the information in a reduction graph by applying a filter to the graph. An example of this type of filtering is given in Figure 2, which shows both an unfiltered and a filtered view of the graph representing the pair

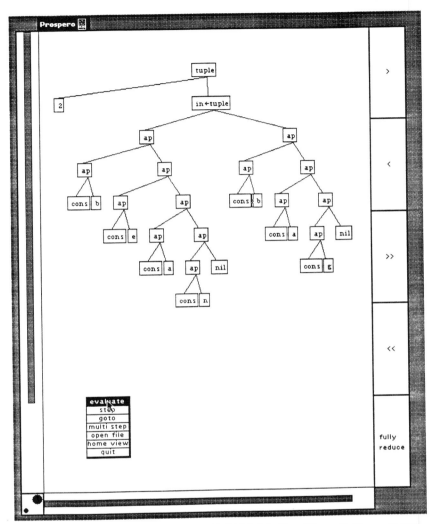

**Figure 2a:** Expression (['b','e','a','n'],['b','a','g']) with an "all-pass" filter

(['b','e','a','n'],['b','a','g']). In this example the detailed information about the reduction graph representation of lists and tuples has been removed.

Filters are functions which take displayable representations of graphs and return a different, often simplified, displayable representation. New filters may be defined by composing existing filters and/or by defining a new

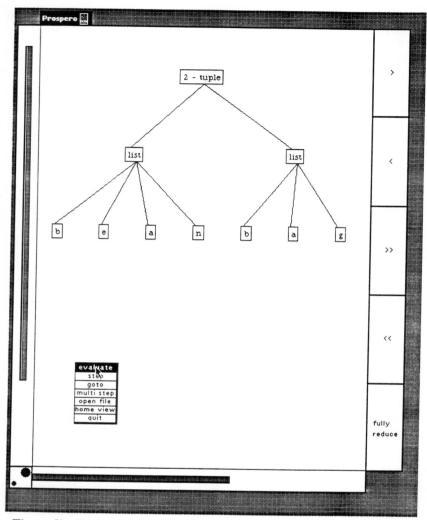

**Figure 2b:** The expression (['b','e','a','n'],['b','a','g']) after filtering using a tuple and list filter

filter function. The filtering illustrated in Figure 2 is achieved by defining and composing three separate filters. The first simplifies the representation of lists, the second simplifies the representation of tuples, and the third is an all-pass filter which can be thought of as an identity filter: it represents all graph nodes as themselves. Prospero offers two other ways of defining filters, described below.

## Filters for Teaching of General Lazy Functional Programming

When teaching general functional programming, it is unnecessary to discuss the low level implementation or mathematical foundations of the language. Expression rewriting will therefore be carried out at a suitably high level of abstraction. Usually this level will be the textual level at which expressions can be defined in Miranda. So, in the example

```
area x = square x
square x = x * x
```

The evaluation of "area 6" can be rewritten using the following steps

```
area 6 => square 6
        =>  6 * 6
        =>  36
```

At no point did we use any information which is at a lower level than the text of the definitions. By contrast, we could have rewritten "area 6" as follows:

```
area 6 => (lx.square x) 6
        => square 6
        => (ly,y*y) 6
        => 6 * 6
        => 36
```

Prospero allows a user to create filters to represent evaluation at a level equivalent to the text of the definitions. An example of this is shown in Figure 3, where the representation of the evaluation has been lifted to a level equivalent to the first evaluation above.

It is also possible for filters to be applied to a reduction graph allowing most of the information below source level to be removed, but leaving some information, say the mechanics of pattern matching, in the display. Of course such a pattern matching filter need not use the lambda calculus notation of the reduction graph, but may present pattern matching in a way which is felt to be more suitable for the students using Prospero.The filter used for these graphs is one which presents expressions at a level of abstraction equal to that of Miranda source. The filter also represents application using juxtaposition.

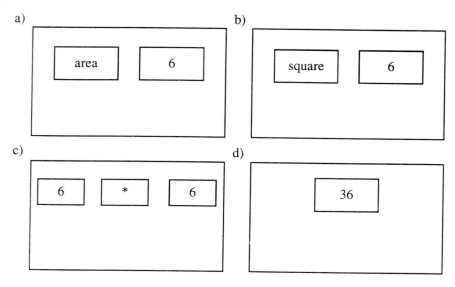

**Figure 3: The four stages in the evaluation of the expression "area 6"**

## *Space and Time Performance*

When teaching programming in any language, it is essential to show how algorithms differ with respect to their space and time efficiency. Both the declarative nature of functional languages and the unusual nature of lazy evaluation make it very difficult for students to reason about and understand how their functions are behaving. Peyton-Jones [5] was referring to experienced functional programmers when he wrote "(i) Seemingly innocuous (and meaning preserving) changes to a functional program may have dramatic effects on its run-time behaviour. ... (vi) It is very difficult to tell when undesirable behaviour is taking place, except that the program runs slower than expected."

In fact, it is very easy to write function definitions which appear to be clear and simple ways of expressing an algorithm, but which turn out to be highly inefficient in terms of time and/or space usage. These inefficiencies are, more often than not, independent of the implementation technique used for the lazy language. They may even be independent of the fact that lazy evaluation is being used. For example, the differences in efficiencies of a selection of sorting algorithms are the same whether or not lazy evaluation is being used.

It is clear that the visualisation facilities of Prospero can be used to great effect in these areas. Filters may be used to increase the effectiveness of Prospero by highlighting properties of data structures. For example, a filter could be defined to represent a list as its size rather than as its contents, or a filter could be used to show how balanced a binary tree was. It is in this area that the power of Prospero's filters are realised, and much more experimentation with these facilities is required.

## Non-Termination

Despite the necessarily widespread use of recursion in Miranda, I have found it to be a surprisingly rare event for students to write a set of non-terminating recursion equations. However, locating such errors when they occur can be very difficult in Miranda. A symptom of non-termination in the Miranda system is either printing the error message "BLACK HOLE" or the failure of the system to print anything.

Prospero's display of intermediate expressions means that a user can see that the calculation is not advancing towards a canonical form. I would be the first to admit, however, that sometimes the form of recursion can be sufficiently complex that it may take the user some time to realise that the evaluation being displayed will not terminate. But Prospero is preferable to the lack of feedback given by Miranda.

It is possible that the application of very abstract filters may be of use in this area. For example, a filter could be used to show whether an argument to a function is approaching a value defined as the recursion base case (for example, the length of a list is approaching 0). At present such filters have not been introduced into Prospero.

## Multiple Views and Filters

It is probable that users will have a number of filters that could usefully be used when viewing an evaluation. For example, a user may want to see a syntax level view of an evaluation, together with a low level view of pattern matching in progress and a view which represents all lists as their length. To make this possible, Prospero allows the user to have a number of different views of an evaluation visible at any time and to associate a filter with each view. A user may create or destroy a view at any time, and all views are updated whenever the user moves through the reduction history, unless the user has marked a view as a "snapshot," in which case the view is never updated.

## Navigation Through Evaluation Histories

So far, I have described how the evaluation of an expression can be represented in the Prospero system and how steps in the evaluation can be presented to a user.

At any point in an evaluation, Prospero allows a user to move any number of atomic reduction steps forwards or backwards through the evaluation history. All views currently open on the graph will be updated to show the result of this move to another point in the history. Any point in the evaluation history may be labelled, and the user may then jump to that label.

### NAVIGATING WITH BREAK POINTS

Another way of moving through the reduction history is by specifying break points. Although I have not fully explored the possibilities of break points in Prospero, it is clear that the following types of break points should be provided:

- Continue evaluation until an error occurs.
- Continue evaluation until a certain function is applied to any argument.
- Continue evaluation until a certain function is applied to a specific argument.
- Continue evaluation until the value of a specific expression is used by another expression.

### CHANGING THE DIRECTION OF EVALUATIONS

I originally intended to allow a user to change the direction which a reduction should take; in this way a user would be able to quickly direct the graph reducer towards the expression which the user suspected as containing errors. I have decided that this facility should not be provided; this decision is based on a simple and yet crucial piece of mathematics known as the Church-Rosser Theorems I and II. These theorems highlight an important difference between lazy evaluation (as used in Prospero) and any other evaluation order.

Informally these theorems give the following warning. For any expression there is at most one possible value to which the expression can evaluate, but if you do not carry out lazy evaluation you are not guaranteed to evaluate the expression to that value. This can be simply demonstrated with the following example.

Given the definitions
```
zero x = 0
badValue = 1/0
```
The lazy evaluation of
```
zero badValue
```
will evaluate to 0. However, any evaluation regime which attempts to evaluate the value of badValue will give an error message corresponding to division of a number by zero.

Allowing a user to change the direction of the evaluation of
```
zero badValue
```
may result in the evaluation of badValue and the production of an error value which would never have occurred if lazy evaluation had been adhered to. It is clear, however, that being shown the evaluation of an expression which you believe to be error-free is tedious and distracting. The solution to this problem lies in the "on-the-fly" filtering. Using "on-the-fly" filtering the user is able to remove the error free expression from the display and so be spared the details of the evaluation of the expression.

## *The Abstract Interface*

Prospero is designed to allow portability from one graphical front end to another and holds no information specific to the way in which graph views are displayed. It is necessary, however, for Prospero to have a data structure to represent the content of the display in abstract terms. It is this data structure, along with the collection of filters, which go together to make the abstract interface.

The data structure used to represent the state of the display is called the abstract display. The abstract display is a collection of abstract windows, and each abstract window represents a view onto the reduction graph. An abstract window has a filter associated with it which is used together with the reduction graph to define the contents of the window.

As an expression is evaluated, the contents of each abstract window is updated to reflect the new reduction graph. The abstract interface is defined in such a way as to make such updates as efficient as possible while maintaining independence from any specific graphical interface.

The portability of Prospero has allowed experimentation with two interfaces to date. The first was based upon the VP system described in

Billyard [1]. The second interface is implemented in Smalltalk and is the interface used during development, which has produced the screen dumps shown in this paper.

*Summary*

The history of the evaluation of an expression is represented as a sequence of reduction graphs. The graphical display of all of the information contained in a reduction graph would make both displaying and understanding the graph difficult. For this reason a user may have a number of views of the graph, each of which may have a filter associated with it. The purpose of a filter is to remove some of the excess information in the reduction graph. A user may also filter out reduction graph detail using on-the-fly filtering or by augmenting the definition of the abstract data types used by the expression. Once users have a view of the reduction graph they may move freely through the evaluation history by stepping through the evaluation or by setting break points.

# Conclusion

Using a simple graph reduction model, Prospero has a representation of each step in the reduction of any expression written in a lazy functional language and a set of rules with which to transform the initial representation of an expression to its canonical form.

Prospero's abstract display provides the user with a number of views of a reduction graph. Each view has a filter associated with it which serves to remove unnecessary information present in the reduction graph. Prospero also provides facilities to allow the user to move freely through the history of an evaluation.

These features can be used to support the rewriting techniques used in the teaching of functional languages. This rewriting technique is usually carried out in a pencil-and-paper manner by students, leading to a number of problems. Prospero provides a tool which allows a student to observe the rewriting of any expression written in a lazy functional language and which allows, through a system of filters, the presentation of a subset of the information present in the reduction graph.

## Acknowledgements

I would like to thank Matthew Huntbach for his valuable comments; Andrew Toal for his advice, constructive criticism and encouragement; Richard Bornat and Keith Clarke for their time and effort as my supervisors; Eliot Miranda for his energy, patience, and Smalltalk expertise; Adam Billyard for his enthusiasm, confidence, and friendship; and Jo 'Beanbag' Hartley for inspiring the examples and everything else.

## References

[1] Billyard, A., Di*rect Manipulation Browsers*. PhD Thesis (in preparation). Queen Mary and Westfield College, London.
[2] Bird, R. an P. Wadler, *Introduction to Functional Programming*. Prentice Hall, 1988.
[3] Mobus, C. and O. Schroder, "Representing Semantic Knowledge with 2-Dimensional Rules in the Domain of Functional Programming," in *Visualization in Computer Interaction*. LNCS 439, 1988.
[4] Pain, H. and A. Bundy, "What Stories Should We Tell Novice Prolog Programmers?" in *AI Programming Environments* (ed. R. Hawley). Ellis Horwood, 1987.
[5] Peyton-Jones, S., *The Implementation of Functional Programming Languages*. Prentice Hall, 1987.
[6] Snyder, R., "Lazy Debugging of Lazy Functional Programs," *New Generation Computing*, Vol 8 (1990), pp. 139-61.
[7] Taylor, J., *Prospero: A System for Representing the Evaluation of Lazy Functions*. Internal Report No. 522. Queen Mary and Westfield College, London, 1991.
[8] Taylor, J., *Presenting the Evaluation of Lazy Functions*. PhD Thesis (in preparation). Queen Mary and Westfield College, London.
[9] Toyn, I., *Exploratory Environments for Functional Programming*. PhD Thesis. University of York, 1987

## Notes

1. Miranda is a trademark of Research Software Limited.
2. The programmers studied were taking a Masters conversion course in Computer Science. Few of them had done any programming before taking the course. All of them had just completed an 11-week course in Modula-2 and Prolog programming.

# CLASSROOM EXPERIENCES

# Computer Assisted Lecturing: One Implementation

## Jacques Raymond

Despite considerable efforts and budgets, the personal computer is not a common tool for professors in their work environment, the classroom, as it is for many other professions. Word processors for secretaries, spreadsheets for many trades, databases for record keeping, accounting packages in many organizations, CAD for designers and architects, and many other PC users do not need any justification, but PCs in the classroom still do.

Despite their shortcomings, the daily tools of a professor continue to be the blackboard, the overhead projector, and the textbook. There are reasons why the PC is not yet central in the classroom.

- The tradition has been the utilization of PCs in the one student-one machine paradigm, not providing a common shared visual space for the class room as a group, as the blackboard does. This would be equivalent to each student reading a textbook in the class.

- Much of the courseware is used to undertake teaching tasks such as drilling and reinforcing, rather than to provide tools for helping professors to teach in a more effective manner.

Many other reasons have been studied [2]; however, we deal only with the two above.

The recent availability of projectors capable of at least VGA-level resolutions and colors has made possible a new role for the PC in the class room. In our approach, the large graphic screen becomes the blackboard, the overhead projector, the slide projector, and its software provides professors with a set of tools designed to effectively support their pedagogical roles in the traditional classroom setting [8].

The tools offered are similar to the visual aids provided by overhead transparencies, which reinforce the professor's role. We, therefore, provide a system allowing the professor to design a course, create the structure of its lectures, draw transparencies in a manner that will reflect the teaching strategy at presentation time, and allow for the manual interventions such as highlighting and marking which are common in presentations. Furthermore, functions to help in the explanations of topics are provided. Since graphics and pictorial information have a large role to play in the teaching of many disciplines, computer graphics techniques and concepts are central in such a tool set.

The PC can now have the role of an assistant helping the professor to draw and write better, to explain and show more effectively, to access references and bibliographic information more quickly, and choose examples from a vast library during lecture preparation and presentation to the class. Few experiments challenge the traditional setting of the PC utilization, although some other approaches exist in different backgrounds and environments [1, 5]. Still, most of the efforts seem to be in the direction of the same paradigm, but with more hardware, more colors, more media, more speed, and more memory.

We presented the Computerized Overhead Projector as one of the early attempts towards our goal [7, 8]. Other examples are listed in [1, 6, 10]. We now present the graphic tools, the design and structuration tools, and the presenting tools that we have developed in software called *EOP* (Editor-Organizer-Presenter).

## Design Stage of a Lecture

### Data Structure Representing a Lecture

The structure of a lecture has been presented by Dick and Carey [3], and we have adapted this structure and its associated vocabulary to our needs. Some well known rules for transparency design (number of lines, amount of information) are explicitly enforced by the editor, while others (single topic, slide unity) can only be implicitly encouraged. Therefore, we view a lecture as a hierarchical structure which contains objectives, transparencies, pedagogical points, and graphic elements.

    1. Course Objective O1
        1.1 Transparency O1T1
            1.1.1 Point O1T1P1
            1.1.2 Point O1T1P2
            1.1.3 Point O1T1P3
            ...

1.2 Transparency O1T2
    1.1.1 Point O1T2P1
    1.1.2 Point O1T2P2
    1.1.3 Point O1T2P3
    ...
1.3 Transparency O1T3
    1.1.2 Point O1T3P1
    1.1.3 Point O1T3P2
    1.1.4 Point O1T3P3
    ...

2. Course Objective O2
    2.1 Transparency O2T1
    ...

An objective is one of the topics covered during a lecture. The transparency level corresponds to its manual equivalent, the acetate, although the flexibilities offered by the PC extend the possibilities to the point where extremes can be easily reached depending on the professor's course design practices. We have seen examples of lectures made of only one long continuous transparency with many dynamic additions and deletions, and examples of a one-hour lecture made of literally dozens of transparencies. Good teaching practice dictates that each transparency should convey one topic. This topic is covered via a series of "points," the set of screen actions which develop the presentation and explanation of that topic. An approximation of the concept of point can be derived from the classic gradual uncovering of an acetate during the unrolling of its presentation, the successive checking of the elements of a bullet list, or the detailed steps of a theorem proof.

The organizer is an outliner whose levels implement the structure defined above. Titles are user defined and, although they are for the instructor's use only, should be meaningful and self-commenting. By appropriately naming each level, the instructor will automatically obtain a transparency list, a course syllabus, or a lesson plan which can be distributed to students. Provided as well is a detailed description of how each topic will be covered. The definition of this teaching strategy is the responsibility of the professor, but it would be very difficult, if not impossible, to draw some graphics in an unstructured manner and expect a structured, logical flow in the presentation. Our software assists the professor in course creation in a much more effective manner than a simple outliner or word processor.

In the actual implementation, each entity (objective, transparency, point) is a reusable object: it can be copied, cut and pasted, or reproduced in another lecture, another objective, or another transparency. The object can also be reused via referencing in other places. References are pointers to the original object, which during the presentation is reused at that particular location. This allows for consistency in figures and diagrams, for the creation of a library of entities, and for a global update of a whole course.

## *The Objectives*

This level simply defines the lecture content and is provided as help to defining the syllabus and the placement of the lecture within that syllabus. An objective is a set of transparencies, and a lecture is a set of objectives. This is defined according to Gagné [4] and Dick and Carey [3] in order to structure the lecture. The functions provided by *EOP* at this level are:

1. At the beginning of each objective during the presentation, *EOP* constructs and displays two slides automatically:
   a- the structure of the lecture with an emphasis on the objective to be covered; and
   b- the structure of the coming objective, as a list of its transparency titles.
2. Objectives are reusable objects and, therefore, can be cut and pasted in other lectures, globally modified, and referenced in libraries.
3. As transparencies are structured under one objective level, it is not possible to manually jump (GOTO?) to a transparency in another objective during a presentation. The transition, if deemed necessary, must be done via branching to the other parent objective first, then to the desired transparency. This enforces structure and logic in the presentation itself and provides some help with the "lost in hyperspace" problem. Note that the same restriction has been applied to the reference bibliography command presented later.

## *The Transparencies*

A transparency is a structured collection of pedagogical points. Although one wished that this would correspond to the way acetates are manually drawn, this structure and the associated scenario make it virtually impossible to draw unrelated items on the same acetate as is so often done manually. Since a scenario has to be created, there is an implicit enforcement of some teaching logic and, consequently, an enforcement of the existence of that logic. Additionally, we provide a scope of new functionalities which extend the teaching possibilities and, therefore, the ways by which a professor might want to treat a given subject.

Transparencies are shown point after point according to the predefined scenario, to the instructor's intervention, or to a combination of both. The presentation may go forward or backward to accommodate questions, repetitions, or clarifications.

Structuring each transparency in pedagogical points serves three purposes:

1. The graphical structure of the transparency reflects the logical structure of the explanation which is represented.

2. The unrolling of the explanation has no geometric link with the final picture being presented. The professor can use the whole screen space, constrained only by the logic of how the topic should be presented. There are no space limits, and this space is not linear as it is on a plastic transparency.

3. For the same reasons, unrolling the explanation has no timing link between its elements. To be sure, a well prepared scenario can be unrolled automatically as in an animation. However, the independence of the objects allows the professor to treat each pedagogical point independently, in any order he or she chooses to accommodate particular questions or situations. The scenario remains the guiding thread for the planned "regular" situation, an uninterrupted lecture.

Overlays are virtually unlimited to any number of transparencies, or even parts of transparencies, in any combination as defined by the course designer. Furthermore, not only can transparencies overlay each other, but they can also overlay, or be overlaid by, the "blackboard," the "slide projector," and the "computer screen," to combine each element into a multimedia image. No difference in treatment is made between these elements, which means that nontextual multimedia images are treated like text elements. They can be cut and pasted, duplicated, reordered, and referenced. Pedagogical points can be skipped, reordered, commented out, or highlighted in many different manners, according to the defined scenario or manual intervention decided by the instructor after class feedback.

Presentation of a transparency is accomplished in three stages. These are designed to mimic the more or less traditional way of showing an acetate:

1. Generally, people show the whole acetate then explain each part, or cover the whole acetate then uncover each part as they explain them. The first step occurs when the acetate is put on the overhead projector. This is the presentation step, that is, what the audience sees of this transparency for the first time. Of course, since the points presented now are logical concepts instead of both logical concepts and graphic items, it is possible to show the initial view of the slide in any visual manner required by the logic of the presentation with no layout limitations.

2. Once the acetate has been put on the projector and seen by the audience, the presenter will treat each point it conveys by pointing, uncovering, or circling all the related parts as a unit. Again the flexibilities are numerous, since parts can be uncovered or hidden, made to blink, to change color, to be framed, or to be underlined. It is the course designer who chooses which scenario is best suited to explain a particular concept.

3. At any time, but usually after the topic has been covered, questions can be accommodated by presenting each or all or subsets of pedagogical points again. This stage is entirely manual since the scenario has been entirely treated, and the presenter is now reacting to audience feedback. At this stage, the flexibilities of the tools available are critical, since we want to offer functions to the presenter to give a better explanation on some topic that was not well understood via the prepared explanation.

## *The Pedagogical Points*

Although logically the same structure as its parent levels, this is the most important level from the pedagogical point of view. The software treats it, however, as any other branch in the course tree. We saw in the previous paragraph that a transparency is a set of these pedagogical points. Each point is made up of graphic elements representing its visual content, scenario commands representing its behavior on the screen during presentation, explanation, and question answering, pointers to link them to "hypermedia assistants," for example, causing the execution of a program, the display of an image, or the start of an appropriate I/O device, and finally its relative position in the structure of the transparency. Graphic elements include the standard text, sketch, image, drawing, geometric object, and macro object. As they are regular objects, they can be cut and pasted, or referenced in libraries. All the graphic elements which make up one point inherit the scenario attribute of their parent. This means that whatever display function

# Computer Assisted Lecturing: One Implementation 167

is applied to the parent, they will behave together in the same fashion during the presentation. They will move, appear, disappear, or blink together. Points represent the steps in the development and unfolding of a description, a proof, or an explanation. The associated scenario and the relative position of each point in the structure of the transparency are a direct software representation of the teaching strategy of the instructor.

Since the software representation is flexible, the text and graphics of the transparencies can be edited at any time for course updating. Since the teaching strategy is represented in the same manner, it too can be updated, edited and adapted by various lecturers of a same course. Changing a method of presentation can be done without necessarily redrawing the graphics of the transparency. Stepwise refinement of a presentation is easily available via promoting elements to the status of pedagogical points. This is a tremendous time saver for a professor once the course has been created.

## *Types of Objects and Object Attributes*

The Lecture is the topmost object. It contains objectives and its attributes define if the presentation is to provide the syllabus automatically at each new objective and if the timing functions are to be enabled. Other attributes include total estimated time, current objective, and transparency.

The Objective contains transparencies and an attribute specifying if this objective is to be included in the syllabus. This is useful because some objectives may be put in a presentation only for reference purposes or as repetition from the previous week.

The Transparency contains pedagogical points and the following attributes:

- estimated time,
- background color, and
- overlay previous screen or erase screen.

The Point contains graphic elements and the scenario commands which are its two attributes. Some hypermedia oriented points (programs or external libraries) may have more specialized attributes. For example, if a slide is to be displayed, the palette and graphic mode must be specified.

The graphic elements which make up a point have one or more of the following attributes:

- type: text, ellipse, sketch, window, and image,
- color: according to the hardware used,
- layout: position on the screen,
- style: for example: fonts or shading, and
- variation: for example, transparent or opaque windows.

In addition to the organizer, a structured graphic editor is provided for drawing, writing, and otherwise specifying the graphical content of the transparencies. The result is not a set of bit-map images representing the graphical layout of the transparencies, but a structured data base of objects representing both the graphical design of the transparency and the pedagogical structure of this transparency.

Figure 1 shows the graph representing a sample lecture, and Color Plates 12–16 present some transparencies, several of which are overwritten by the blackboard during an actual presentation. The incomplete images in some figures attempt to convey the animated aspects of their presentation.

## The Presentation Stage

### The Scenario

The presentation and highlighting commands are the point attributes which control the presentation. These commands are also available manually for direct control of the presentation. Entering a command manually will override the scenario, thereby providing ways for the presenter to modify the ways things are shown. Many effects are offered: show, invert, erase, animate, displace, underline, change color, change palette, frame, cross out, make an appearance, load, run, and more.

### The Unrolling

In addition to the manual override of the scenario commands, the timing and order of the transparencies and pedagogical points can also be modified as follows:

- move on, the "normal" command;
- backup one given level of presentation;

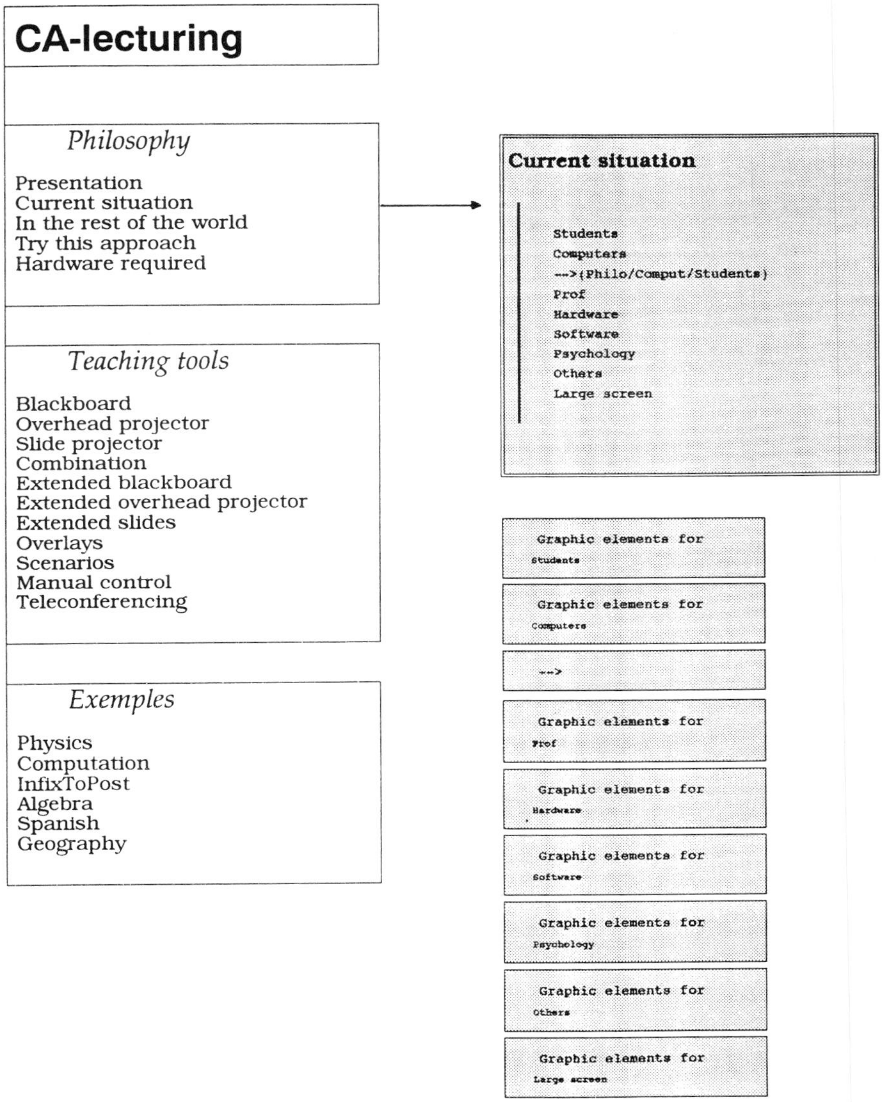

# Computer Assisted Lecturing: One Implementation 169

## CA-lecturing

### *Philosophy*

Presentation
Current situation
In the rest of the world
Try this approach
Hardware required

### *Teaching tools*

Blackboard
Overhead projector
Slide projector
Combination
Extended blackboard
Extended overhead projector
Extended slides
Overlays
Scenarios
Manual control
Teleconferencing

### *Exemples*

Physics
Computation
InfixToPost
Algebra
Spanish
Geography

---

**Current situation**

    Students
    Computers
    --->{Philo/Comput/Students}
    Prof
    Hardware
    Software
    Psychology
    Others
    Large screen

Graphic elements for
Students

Graphic elements for
Computers

--->

Graphic elements for
Prof

Graphic elements for
Hardware

Graphic elements for
Software

Graphic elements for
Psychology

Graphic elements for
Others

Graphic elements for
Large screen

**Figure 1: Course graph**

- re-present a given level in the presentation;
- abort a given level of presentation;
- skip next level of presentation;
- move to a specific transparency, via a specific objective;
- consult reference;
- replay scenario; and
- replay explanation.

*Manual Entries*

Another set of tools in the blackboard mode allow for manual entry (sketching, notes, or overwriting). These entries can be made via different services depending on the actual configuration. Keyboard, mouse, graphic tablet, scanner, and electronic pressure sensitive chalkboard are now supported. The blackboard includes colored "chalks" and color "erasers" for color-selective erasing. The manual entries can be erased while leaving the overwritten slide intact; the palette can be manipulated to show components in a scanned image. Erasing the board can be done globally, by points, by color, or by erasing the manual entries but not the slide, allowing the presenter to focus the attention of the audience on whatever parts of the transparency the presenter wishes.

*Timing*

Timing tools are also offered to help the instructor. They include a chronometer for timing exercises, a visual feedback to indicate if the lecture is going too fast or too slow, a wall clock, time since start of lecture, time remaining, and specifiable warnings.

*Other Functions*

1. Use of other software generated screens.
2. Running programs: A scenario command will cause the execution of any MS/DOS compatible program (memory size permitting). This allows the presentation of independently written software within a presentation that can be used by the instructor to illustrate or complement a topic. The resulting screen can be overlaid by the rest of the transparency.

3. User provided scenario commands: programs can be written for specific purposes that can be interfaced with the course structure. The transparency data is passed along with the necessary attributes in a well defined data interface, making *EOP* an open ended system.

### *Teleconferencing*

Since echoing keystrokes and mouse movements to a modem are easily done, the whole presentation of a course can be done via teleconferencing. One *EOP* is set in the PROF mode, the others in the STUDENT mode. The course and its elements are downloaded ahead of the lecture. The telephone-line traffic consists only in the keystrokes and mouse movements entered by the presenter (who controls all the PCs, connected in a star configuration via a bridge). Switching functions and a communication protocol are provided to allow the presenter to pass the floor to a student in a remote site for fully interactive, but disciplined dialog capability.

## Conclusions

We believe that the *EOP* tools we have developed provide a solution to the two problems presented at the beginning of this paper. Our experience with the use of these tools in the classroom has been very positive. It remains to be seen, however, what the results of a test on a wide spectrum of disciplines would be.

*EOP* provides and improves upon all the functions of

- the blackboard: writing, drawing, erasing, while adding new functions such as selective erase, selective blink, overwriting board, slides, and pictures;
- the overhead projector, to which we added animations, unlimited overlays, hiding any part of a slide or groups of related parts, and more;
- the slide carousel, now with random access, palette manipulating, pointing, and overwriting;
- the computer, providing a medium to run software prepared independently, incorporating their screen output into the common *EOP* visual space and allowing highlights, pointers, and overlays; and
- the textbook, offering a new medium for course creation, presentation, and distribution. This new "textbook" has the advantage of being user

modifiable. The organizer can be used to shuffle the order of presentation, the editor to add personal slides, notes, and touches, and the presenter to insist on topics more important for a particular environment. Students can rerun the presentation at their own speed for reviewing purposes, and if necessary make their own in project presentations, using a common library of icons, functions, slides, and exercises.

## References

[1] Balkovich, E., S. Lerman S., and R. P. Parmelee, "Computing in Higher Education: The Athena Experience," *Communications of the ACM*, 28:11 (November 1985), 1214–1224.

[2] Chambers, J. A. and J. W. Sprecher, "Computer Assisted Instruction: Current Trends and Critical Issues," *Communications of the ACM*, 23:6 (1980), 332–342.

[3] Dick, W. and L. Carey, *The Systematic Design of Instruction*. Scott, Foresman and Company, Glenview, Illinois, 1985.

[4] Gagné, R., *Principles of Instructional Design*. Holt, Rinehart and Winston, 1988.

[5] Hativa, N., "Designing Flexible Software for the Electronic Board," *AEDS Journal*, (Fall,1984), 51–62.

[6] Kolitsky, M. A., "The Electronic Blackboard: A New Way of Teaching Biology," *Western Educational Computing Conference*, 36–40, 1985.

[7] Raymond, J., "Le tableau noir electronique," *5th annual conference on teaching and learning*, Ottawa, June 1985.

[8] Raymond, J., "The Computerized Overhead Projector," *Computers and Education*, 11:3 (1987), 181–195.

[9] Raymond, J. and Y. Toussaint, "Un protocole de communications pour les cours par téléconférence," *Département d'Informatique, Université d'Ottawa*. Rapport Technique. Mars 1988.

[10] Stefik, M., G. Foster, D. Bobrow, K. Kahn, S. Lanning, and L. Suchman, "Beyond the Chalkboard: Computer Support for Collaboration and Problem Solving in Meetings," *Communications of the ACM*, 30:1 (January, 1987), 32–47.

# Interactive Computer Graphics via Telecommunications

## Joan Truckenbrod
## Barbara Mones-Hattal

Telecommunications connects people in different geographic locations and enables them to interact with one another as if they occupied the same space. Using telecommunications we create a virtual reality involving people from around the globe simultaneously. This has significant applications in the arts, education, science, and industry since it provides a real-time, interactive link between people and projects in distant geographic locations. Telecommunications is unique in relation to other forms of communications as it involves visual images, diagrams, text, voice, music, sound effects, and animation, as well as interactive capabilities. Integrating a computer graphics system into telecommunications creates a multi-media network that encircles the earth, weaving together a diverse set of participants. These new avenues of communication have dramatically altered the ways in which we perceive the world and process information. They rely upon some sophisticated technologies that provide a wealth of options for us — individually and collaboratively — to gather and manipulate information, and results in the creativity spawned from synergy.

## Telecommunication Methods: Current Technology

Telecommunications has valuable applications in education, business, recreation, and in the art and design professions. Currently there are four methods of developing telecommunications processes and projects, reflecting levels of accessibility and cost. The most dynamic of these methods is *live, full-motion video teleconferencing*. This process facilitates live, real-time interaction and discussion between two or more groups in different geographic locations. Live video images or pictures of each group are recorded by a series of video cameras and the video signal is are transmitted

Figure 1: A video teleconference

to other groups throughout the world. Thus each group sees the other on a large video display adjacent to a video display of their own group. Each teleconferencing facility has two video displays or monitors, one for the incoming video signal carrying images of the other group and one for the outgoing video signal carrying the images of the local group. The fact that these video screens are adjacent to one another at each site suggests that interesting interactive performances occur involving both groups simultaneously.

An example of this type of display is shown in Figure 1. This is a photograph of one of the first teleconferences created by artists who explored the nature and character of the teleconferencing process and began to examine the vocabulary of this media for artistic expression. This took place in 1987 involving Tom Klinkowstein, Rob Fisher, and their students on the east coast and Joan Truckenbrod with students from The School of the Art Institute in Chicago. The Chicago group was joined by Werner Hertrich from the Art Institute, John Grimes from Illinois Institute of Technology, and Joel Slayton from San Jose State University. This project was supported in part by the SPRINT company

Computer graphics, computer art, and animation can be incorporated into any teleconference since computer graphics systems are easily integrated into the teleconferencing process. Teleconferences are multi-media events involving text, voice, music, photographs, diagrams, sketches, and paintings. Because participants can, and are encouraged to, respond immediately, teleconferencing is a highly interactive process. In addition, participants in different locations can work on the same project using computer software called Timbuktu for the Macintosh. This allows people in different locations to work on the same document, diagram, or drawing together using the mouse; people in different places can both work on the same Macintosh screen image at the same time. This facilitates collaborative projects between educators and students, between artists, or between artists and musicians in different parts of the country or the world.

Live teleconferencing requires facilities which are available in many universities and businesses. Transmission of live video and sound can be accommodated through satellite transmission or fiber optic cables. Using satellite transmission requires a satellite uplink and downlink with the appropriate connections to the conferencing room. Fiber-optic transmission requires at least a T1 line to facilitate the real time transmission of live video and sound. The quality of the video transmission is variable depending upon the level of fiber optic cable available. The cost of live video teleconferencing is very high. Facilities for this type of conferencing can be rented through communication companies such as AT&T, Ameritech, and SPRINT.

A second and much more accessible type of teleconferencing uses video phones to exchange pictures and conversation over ordinary telephone lines. Video phones transmit a video picture in a slow-scan mode in which photographs are transmitted line by line. This process is similar to the way in which pictures from space were sent back to Earth from the early satellites and broadcast on television. Video phones have small video cameras in them, but they can also freeze and send pictures from other video cameras, video tape, or television. Black-and-white video phones are available from Sony Corporation. A color video phone, or slow scan device called The Robot, is available from Robot Research in San Diego, California. Using these devices, pictures of any artwork, environment, or group of people can be transmitted anywhere in the world as long as the sending and receiving groups have compatible video phones. When using video phones it is valuable to use two phone lines to accommodate the transmission of voice on one line while pictures are being transmitted on the other line simultaneously. When using only one phone line, the participants must alternate the transmission of voice and pictures, as only one transmission is possible at a time.

Computers can also be integrated into this process if they have a video-out signal that can go into the video phone. The use of a computer graphics

**Figure 2: The Electronic Cafe**

system adds a dynamic component to this process as incoming images can be captured (digitized), processed, and sent back. For example, images from both locations can be combined or artists from two locations can sequentially add drawings to the composition as it is sent back and forth.

The advantage of video phones is that teleconferencing is possible anywhere in the world where there is access to a phone line. Images from video phone teleconferencing have a unique visual quality as are shown in the black-and-white figures from a New Year's Eve "Round the World" link sponsored by Sherrie Rabinowitz and Kit Galloway at The Electronic Cafe in Santa Monica, California. Video phone teleconferences are created with artists in each time zone around the globe, at midnight. Figures 2 and 3 are from The Electronic Cafe and Joan Truckenbrod's Studio respectively. This is a highly interactive process, particularly in a New Year's Eve party atmosphere.

The third method for interactive communication with people in a distant geographic location is through the use of a modem connected to a computer and to telephone lines. Documents, diagrams, drawings, and pictures can be sent between computers over phone lines. A common use of this process is for electronic mail. In real time, however, modems can be used for a dynamic exchange of ideas and information. There are many potential applications of "modem dialogues" in education, the arts, and business. An interactive art performance is created by developing a collaborative art project by sequentially adding, subtracting, and transforming

# Interactive Computer Graphics via Telecommunications

Figure 3: Joan Truckenbrod's studio

elements in the visual document. Text and photographs are integrated into a collaborative creative process. In education, modems allow a lively exchange of ideas. Teachers can also work with students in different locations using modems and computers, allowing them to work with ill or handicapped students who are not able to attend classes, and allowing students to work with teachers with specialized expertise at different locations. Robert Crago, from Australia, has developed software called "Electronic Classroom" that allows a teacher to work with five students at different locations simultaneously. Using this software, students receive lessons and have their projects reviewed by the teacher over the modem. The teacher can view the learning process of each student from his or her computer. In addition, each student can share his or her work with the other students over the modem. This software allows teachers to work with special students at other locations and to make unique courses available to interested students who are not in the same geographic location. Transmission rates currently range from 300 baud to 9600 baud, with 9600 baud preferred for picture transmission. Modems are available that will recognize the speed of the incoming information and will adjust automatically to that speed. In addition, modems must accommodate standard telephone formats. The United States has the Bell standard while the international standard is CCITT.

The fourth method of telecommunication involves private closed circuit networks, or local area networks (LANs). LANs allow similar options for viewing, sharing, and transfering files, but computers are located in close proximity to one another. The transfer of the files is accomplished directly via cable network and does not require phone-line transmission. This allow groups of people to work on projects in the same general area, in a laboratory environment, and still benefit from the advantages of a shared computer desktop.

## Expanding the Educational Process Using Telecommunications

Telecommunication can enhance opportunities for the educational process in a variety of fields. Students in one location can receive a lecture from an expert in another city. A course can be created involving lectures and discussions with professors throughout the world. Specialists can be called in to critique projects using telecommunication facilities. Students and professionals can collaborate on projects even though they are at different-research or educational facilities. Even in the same location, students and faculty can communicate using LANs such as Appletalk for the Macintosh computer.

A number of models can be built for this educational process. Initially, a course can be taught in one location by a professor in another part of the world. Another approach is to involve experts throughout the world as speakers in a course. A seminar environment can be created to stimulate a dialogue among students and faculty located throughout the world. Telecommunications can be a two-way interaction involving students and/or faculty in two different locations. In addition, a telecommunication "bridge" can be created to involve an entire small group of participants in different geographic locations simultaneously. A third mode of telecommunication is similar to broadcast television, in which a lecture series or exhibition series is broadcast via satellite and can be received by many audiences. The reception of this type of broadcast requires satellite down-link capabilities. Thus, with live full motion teleconferencing, video-phone transmission, modems connected to computers and telephone lines, and private, closed-circuit networks, there are four methods of utilizing telecommunication facilities for teaching courses, holding seminars, presenting lecture series, and creating exhibitions of computer graphics and video art and animation.

## Case Study — The National Geographic Kids Network

Utilizing telecommunications, some particularly interesting projects are emerging in the classroom for young people. The National Geographic Kids Network, developed by the Technical Education Research Center (TERC) with funding from from the National Science Foundation, gives students the opportunity to collect and analyze data and to share their findings with one another through extensive use of telecommunications.

TERC has developed telecommuncations software for students in fourth, fifth, and sixth grade science classes for work online in exchanging data, comparing observations, and testing hypotheses on subjects they explore in common. After collecting the data, students from these class "research teams" send the raw data to the Kids Network Staff and a "map file" that provides a geographic overview of the national data is sent back. The students can then see on a map (a graphic) how their statistics compared to their fellow students and to the collective data in the United States or in other countries. Dr. Candace Julyan is the director of curriculum and training for this project. She works closely with the two hundred schools participating in the field tests of the materials. Thus far, the topics have centered on global ecology. Dr. Julyan stresses that "the approach of the Global Laboratory Project emphasizes a critical, yet often missing part of science education, that of direct participation by students in the conduct of science. We want to further develop a model for learning science that is project-oriented and interdisciplinary, that emphasizes depth of knowledge, and relates to the needs of society" [7].

## Case Study — Course in Telecommunication Arts, The School of the Art Institute of Chicago

Joan Truckenbrod recently developed a course in Telecommunication Arts at The School of the Art Institute of Chicago. The objective of this course is to explore various transmission technologies and to study the potential of integrating telecommunication technology into the creative process. Can interactive, real-time communication processes provide a vehicle for artistic expression? Based on the work in this course, it is clear that telecommunication technology integrated with computers provides a unique experience for art students, stimulating creative thinking that expands the horizon of artistic expression. The course includes a history of artists using telecommunication technology with discussions of various telecommunication events.

Sherrie Rabinowitz and Kit Galloway presented lectures to the students in Chicago from the Electronic Cafe in Santa Monica, California using video phones. They discussed projects that they had initiated and showed examples of different telecommunication events that they had sponsored. Students were introduced to the wide range of telecommunication events that are possible in the context of teleconferencing. During the semester, students experimented with integrating computer graphics into the telecommunication process with the use of imagery and text. Image processing techniques available with the Robot device were also explored and integrated into the artwork.

Students created an interactive performance using both color and black and white video phones with the Electronic Cafe. Color Plate 17 is a composite image from the Electronic Cafe using the Robot device. They choreographed sound, music, voice, and a sequence of imagery and text together for this performance. In addition they transformed pictures and conversation received from other participants in real time and transmitted them back to the participants. The processed image in Color Plate 18 is an example of this type of transmission. The projects in this course illustrate the viability of live interactive performance art involving participants from around the world. In addition, telecommunications is a valuable mode of interactive instruction in the educational process.

A graduate student at The School of the Art Institute works with popular culture icons that she places in public spaces. She uses common objects to create a new awareness of these iconic images. As she is working with icons from computers and electronic technology, it is natural for her to send these images into "electronic space." She is currently developing a project to be implemented on one of the electronic bulletin boards.

## *Case Study — Local Area and Long Distance Networks in Studio Art, George Mason University*

New approaches to collaborative and interdisciplinary design environments are utilizing methods of telecommunication in order to support the computer imaging curricula at George Mason University. Current projects involve local area networked design environments as well as the long distance transmission of files over electronic networks. In many cases the telecommunication method has contributed only one part to the overall design process but in every case this part has been extremely important to the process and resultant product.

# Interactive Computer Graphics via Telecommunications

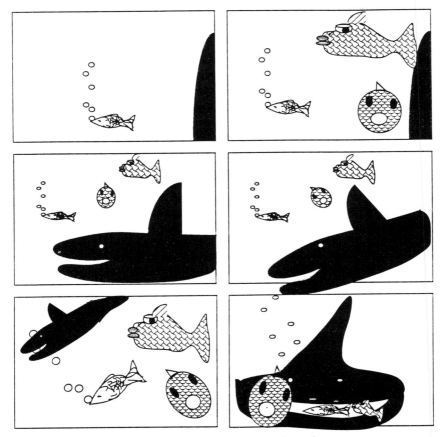

Figure 4: The initial storyboard

Timbuktu, contributed by Farallon Software Company, has been used extensively in the undergraduate computer graphics curriculum to introduce new software products and demonstrate software tools. Most importantly, it has been used to introduce students to collaborative design projects using computers. Timbuktu encourages real time collaboration by allowing many guest users to connect to a host and share his or her screen with all of the other participants. Users can make changes to the host's document and control of the software can be passed between the host and the guests. In the teaching laboratory students can share their work with other students by allowing them to draw on their workspace or simply view their work while they demonstrate the use of a software tool.

Figure 4 is the initial co-designed storyboard for a project in a first semester, computer graphics course for art students. This attempt was accomplished after an initial session of approximately one hour using Macintosh II computers, Timbuktu, and MacDraw. During this time, students were acclimating to a relatively untapped environment for artists that allows for joint control of the drawing, painting, or sculpting tool. This computing environment encourages several students to draw in the same picture plane simultaneously. This method of developing ideas is often overlooked in studio art environments that almost always exclusively encourage individual expression. Figure 5 represents the students' second storyboard as a result of their second session together.

The first storyboard developed into an underwater environment in which the smaller fish are eaten by a large shark. During the drawing session, the students discovered distinct aspects of their individual style in the development of their work together. For instance, Michelle Kim's drawing style is representational; Paula Croisetiere's is cartoonlike. Figure 5 represents a recognition and synthesis of these differing styles. One fish is now realistic, and one is drawn as a cartoon character. Now, in the updated storyboard, a television sinks to the bottom of the water. Paula's fish swims out of the cartoon environment in the television and swims into the ocean. Michelle's representational fish swims from the ocean into the television. The storyboard has evolved interestingly from the design issues that developed in the first session. Each student's drawing style is represented in the final storyboard, and in some places there has even been a seamless merging of the two styles, places where it is unclear which student was drawing. Color Plates 19 to 27 follow the development of the jointly drawn and painted sequence.

There are, however, problems that emerge in a collaborative design environment. The students sometimes don't use the recognition of the other students' different approaches and styles as potential for creative material with which to work. Also, there is often an uneven commitment to the problem solving involved in the project. One student sometimes works harder than another. There are numerous potential work-group pitfalls. The best collaborative teams require strong commitments from all members. Abilities to coordinate schedules, workloads, and exchange technical information are part of the evaluation procedure for these projects. In addition, students using Timbuktu can communicate extensively with each other and rely less on the instructor as the "all knowing" information provider. This can alter the role of the instructor. The instructor can take on a supportive or

# Interactive Computer Graphics via Telecommunications 183

Figure 5: The second version of the storyboard

consultative role at certain times during the course of a semester and allow the students time to help each other.

Another undergraduate project that introduces students to three-dimensional modeling utilizing electronic networks is represented in Color Plates 28 to 32. This project pairs two students in order to pose questions about visual communication skills. One set of students is asked to go to a distant location on campus and describe a meeting room. They can describe the room by measuring doors, walls, etc., and photographing interior views as well as views outside any windows. This information is networked to another set of students in the lab whose job is to reconstruct the space in a three dimensional computer modeling program (Topas from AT&T) from just the verbal and visual description they are receiving. The students are then reunited; work on the final construction is performed in the lab with students pairing off and discussing the successes or failures inherent in their attempts. Both sets of students then go to the space and "fix" their models to present the "real" space as accurately as possible. Students quickly understand this room in detail, having analyzed and scrutinized most of it. At this time, students pair off and alter aspects of the room in the computer model to make an expressive statement about the relationship of objects in our real world to simulated computer spaces. Often text is introduced to support an idea. Models are rendered in stereo pairs, installed in the "real" location and viewed with Franka-type stereo viewers. Color Plate 28 is the view out the windows of the room. Color Plate 29 is an interior photograph of the meeting room. Color Plates 30 and 31 are computer-generated versions of the same space with the students alterations. Color Plate 30 represents a transformation of the view outside one picture window to that of a Greek courtyard. Color Plate 31 represents the transformation of the computer space to include an interior aquarium, with fish swimming out of the tank and into the room toward the viewer. Color Plate 32 is an informal view of the final installation of the stereo viewers in the physical space.

As part of the Fall, 1991 graduate curriculum, art students at George Mason University in Fairfax, Virginia will be working with engineering design students at the Colorado School of Mines via telecommunication networks. Students at George Mason will be enrolled in a new course called "Collaboration in Computer Graphics" under the direction of Barbara Mones-Hattal and in Colorado the engineering students will be enrolled in "Advanced Engineering Design" under the direction of Michael McGrath. Two projects are planned for the next academic year with the goal of investigating the curricular advantages of pairing students in different locations and areas of study on the same project. One project involves the design and construction of computer-generated sculpture. The art students will be designing site-specific sculpture for an installation. Using CAD software for their designs, the students will transfer the files via network to a "network pal" in Colorado. The network pal will receive the file, output it to paper or a milling machine, make suggestions about the structure of the object, and send it back. This process will continue until both students, at each site, agree that the artistic-design process and the engineering-design process have been completed. In the meantime, information about the process itself will be gathered and discussed. A copy of the sculpture will then be constructed and placed in its site. The network pals can then decide how many copies can be made from their prototype. The second project will pair students from these classes to approach challenges for a new virtual-reality application. The art students will design and build three-dimensional computer models for virtual environments and the engineering students will design new hardware applications.

These students are learning a collaborative design process that supports and reflects current needs in industry. Students who have good communication skills and work well on collaborative teams may be better prepared for their future employment. Thought processes and ultimately the understanding of the design process and product will be positively affected by these experiences.

## The Interactive Nature of Telecommunication Art

As has been described above, computers are being used interactively in telecommunication events to create text, images, and sound, and to transform imagery and sound. Telecommunication processes facilitate an interactive art form that is a true collaboration between people in distant geographic locations, involving visual imagery, sound, voice, music, and text simultaneously. Artists experiment with choreographing telecommunication events in which colleagues from around the world may interact through a visual and verbal dialogue in a hypothetical electronic space or "telespace." This telespace becomes the stage for a new genre of interactive, real-time collaborative performances involving artists around the world.

Collaborative performance is an area that showcases the interactive nature of telecommunication. Faculty of the Institute of the Arts at George Mason University have begun to investigate the role of computer imaging in the design process as well as the presentation of performance art. Equipment in the department of Art and Art History will be utilized in conjunction with collaborative meetings with set designers, musicians, dancers, directors, and animators in order to develop plans, simulate environments, and design appropriate animated sequences for contemporary performance.

## The Role of the Educator, the Student, and the Artist

The future of computer graphics in education is to expand beyond the lab or studio and interface with resources throughout the world. Telecommunication provides the capability to expand the educational arena to a "global learning lab" that includes a whole range of unique resources that enhance the learning process. This process facilitates world-wide access to ideas and information. Computer graphics are an essential component of teleconferencing because they provide visualizations, simulations, and pictures that can be transmitted anywhere. More importantly, however, is the ability to process, modify, transform, and expand ideas visually, in real time, collaboratively with others.

With telecommunications capabilities, educators could maintain flexible schedules and work simultaneously in more than one geographic location. They could pursue unique and broadly based research and develop more interactive and effective teaching and learning initiatives. The use of this technology in educational settings could have dramatic effects on student-teacher relationships. Theoretically, teachers could spend less time writing and delivering lectures and more time assisting students individually in their studies. Visual resources for instructional support would be easily accessible and interactive feedback devices for students could encourage better communication and potentially compensate for reticent student response.

Students could work with any number of professors throughout the semester, analyzing problems from a unique combination of perspectives, and formulating new theories with the aid of collaborative computer graphics via telecommunications. They could be paired with students of differing expertise and together pursue interdisciplinary projects. As potential examples, art students and music students might collaborate on the design of an animated sequence or art students and biology students on a scientific visualization.

Through this approach, artists can now explore alternative forms of expression and simulate many more potential answers during the design time. They can extend the time for constructive play and exploration and enhance the quality of design processes. Artwork may be networked to distant locations in order to be output on similar, or much more unusual, and unique devices.

## Looking Toward the Future

Current technology has two drawbacks: access and cost. With only one or two fiber optic rings available per city and few lines to those rings, few artists or educators can gain easy access (or any access at all) to the technology. One hour of teleconferencing costs approximately one thousand dollars for satellite transmission. This is not encouraging news for those of us who would like to experiment.

Also, there are many concerns about the use of any new technology in the classroom environment. Too many sources of information, if not ordered, could be confusing for certain students, and collaboration among students and/or faculty requires careful organization and planning. Communications skills between these people will influence the success of any project. This becomes exponentially important in transcultural environments.

We hope that this situation will change for the better as the technology becomes more sophisticated and interactive multimedia via networks becomes commonplace. In the meantime, opportunities for the development of better technologies and more effective teaching environments within educational institutions offer a vast array of untapped potential.

### *References*

[1] Carl, David L., "Electronic Distance Learning: Positives Outweigh Negatives," *T.H.E. Journal*, 18:10 (May, 1991), 67–70.
[2] Davis, Douglas and Allison Simmons, eds., *The New Television: A Public/Private Art*. MIT Press, 1977.
[3] Eco, Umberto, *Travels in HyperReality*. Harcourt Brace Jovanovich, 1990 (originally 1976.)
[4] Elmore, Garland, "Planning and Developing a Multimedia Learning Environment," *T.H.E. Journal*, 18:4 (February,1991), 83–88.
[5] Grundman, Heidi, ed., *Art Telecommunication*. Contemporary Art Press.
[6] Inglis, Fred, *Media Theory*. Cambridge, MA: Basil Blackwell, 1990.

[7] Julyan, Dr. Candace L., "National Geographic Kids Network: Real Science in the Elementary Classroom," *Classroom Computer Learning*, 10:2 (October, 1989).
[8] Julyan, Dr. Candace L., "Messing about in science: participation not memorization," *National Academy of Sciences*, October 1988.
[9] McLuhan, Marshall, *Understanding Media*. London: ARK Paperbacks, 1987 (originally 1964.)
[10] McLuhan, Marshall, *The Gutenberg Galaxy*. University of Toronto Press, 1988 (originally 1962.)
[11] Ronell, Avital, *The Telephone Book: Technology, Schizophrenia, Electric Speech*. University of Nebraska Press, 1989.

# Collaborative Computer Graphics Education

## Donna J. Cox

### Collaboration and the Cult of the Individual

During the Renaissance, many scientists and artists collaborated and published works together. Books were created as joint efforts among the biological and anatomical researchers. The scientists worked in tandem with the artists who carefully recorded in great visual detail countless taxonomic studies (Ronan [34]). Renaissance philosophy encouraged the visual study of nature, and many people believed that this visual study could reveal the hidden laws of nature. Biological and anatomical illustrations were invented as a valuable tool that is still used today by physicians and scientists. Artists and scientists were given equal credit in the books that they generated.

The visual study of nature provided insight into worlds that had been mysteries before and set the stage for the scientific revolution. Scientific methodology inscribed the process of the visual, objective gathering of data and the formation of hypotheses and theories. Eventually, famous scientists such as Galileo and Newton became heroes to the scientific community. Out of the age of enlightenment emerged the era of individualism. Paradigm shifts have been attributed to individuals; however, these shifts were ideas whose time had come and were the collective work of many individuals (Kuhn [26]).

Art and science diverged (Weininger [44]). By the turn of the twentieth century, most scientific research was carried out by individuals rather than collaborative teams. Only in 'big' science has there remained a need for researchers to collaborate and pool resources (Smarr [36, 37]). Teamwork and collaboration are extremely important to most of this large-scale science; for example, the space program is a huge, collaborative effort among technologists, artists, and scientists. However, astronauts have received extraordinary publicity as individual heroes compared to the entire team (Baker, [1]).

A similar evolution occurred in art, and great collaborative works became rare. Art from the Renaissance until today has been attributed to the individual artist, and an aversion to collaborative art has developed. Painters and sculptors have adopted 'individualism' and personal expression as the modus operandi (Varnedoe [40], Cox [12], Hamilton [23], Janson [25], Lucie-Smith, [27]). Individualism coupled with novelty have been considered an important characteristic of the avante-garde. Fine art in Western culture is considered a product of individualism, and the fine art world has been preoccupied with the cult of the individual and has had a long-standing bias against collaborative work (Rosler [32], Segard [35]). Academic education in the arts has continued to proliferate personal expression rather than collaboration.

Filmmaking and television production are collaborative art forms. The primary reason for this organized approach is because of the complexity and technological requirements involved in filmmaking or television production (Sobchack [38], Wead [43]). For the same reasons, films and most computer animations are products of a team. Yet a film or computer animation is only considered fine art when the work is close to the personal expression of an individual. The primary reason for computer graphics animations to be collaborative is because of the immense computing power and technical expertise required to implement the work. "But the extensive use of high technology requires collaboration, a process that is in direct opposition to our tradition of creative individualism." (Segard [35], p. 26)

*Now, with the advent of supercomputing and scientific visualization, there is a demonstrated need for collaborative skills to be coupled to computer graphics education. Scientific visualization involves the bringing together of a variety of artistic, technical, and scientific expertises. Many projects involve teamwork since the process of scientific visualization is complex and requires a variety of disciplines (Cox [11, 12, 13, 15], Burson [5, 6], McCormick [28], Freidhoff [20]). Global challenges and environmental issues require collaboration among eclectic social groups to solve important scientific research problems ranging from global warming to stratospheric ozone depletion. Such international problem solving will make strange bedfellows.*

While it is recognized that we are moving toward global economics and environmental economics, we also see a continual emphasis on the individual. "The great unifying theme at the conclusion of the 20th century is the triumph of the individual ... This new era of the individual is happening simultaneously with the new era of globalization" (Naisbitt [30], p. 298–299). However, such a philosophy should not preclude the collaborative process; rather, it should give attention to the collaborative and communication skills of the individual. The individual is empowered through both leadership and teamwork.

## *Building Collaborative Skills*

Historically, the creators of the liberal arts education in the American university system intended to educate students with a variety of disciplines. The British higher educational system forces students to specialize early, concentrating on depth rather than breadth. The American Liberal Arts degree was a reaction to the British system. However, this eclectic system has not been successful in providing a mechanism for collaboration among disciplines. Students specialize and have only a minimal exposure to other disciplines. Nor do undergraduates get a real-world understanding of the importance of collaboration and how teamwork will play an important role in career development. Graduate students are sometimes exposed to collaboration if the dissertation topic lends itself to this modus operandi. Yet, graduate research generally requires the student to isolate rather than pool information with his or her peers from other disciplines. This deficiency in our educational system is a disadvantage to many students who enter the corporate culture where teamwork is considered a priority in the working structure. While there are exceptions to the above and some programs provide interdisciplinary course work (Cox [9]), for the most part computer graphics education, like most education in general, does not provide a mechanism for providing students with collaborative skills.

This deficiency in our educational system is unfortunate since in the computer graphics industry most work is, in fact, collaborative because of the nature of the technology and the structure of many companies associated with the graphics industry. Regardless of whether one is a high-end user of computer graphics applications packages or is involved with software development, teamwork and communication skills are extremely important to career development (Brown [3], Inose [24]).

The following describes an educational experiment to build collaborative skills in computer animation students. First, the author describes the technological environment and class structure, then outlines class projects and results.

## The Renaissance Experimental Laboratory

The Renaissance Lab (REL) was created from a donation grant of twenty Silicon Graphics Personal Irises (PI). In the third year of REL, Silicon Graphics upgraded the PIs to 4D/25 turbo graphics workstations. These workstations are networked via Yellow Pages to a 220 SGI server with 1-gigabyte disk and 32 Mbytes of memory and an SGI 360 Power Series VGX with 64 Mbytes of memory. The Experimental Visual Techniques class is one of five courses in the multidisciplinary REL. Other experimental courses include physics, chemistry, scientific visualization, and descriptive topology. Interdisciplinary education is the focus of REL. This laboratory is a subset of the Numerical Laboratory where NCSA metacomputer networks a garden of architectures ranging from the IBM RS6000 workstations to the Cray YMP supercomputer. However, most educational courses reside locally in the REL, and all of these classes run on SGI hardware since this hardware has proven to be the best for a three-dimensional computer graphics applications environment.

This educational facility is prototypical and interdisciplinary. The Beckman Institute provides an enriching, interdisciplinary environment for NCSA research in advanced computational techniques. REL is a unique educational facility embedded in an advanced scientific and technology research institute. Courses have included university mathematics, computer science, chemistry, and physics curricula as well as certificate programs at a local community college.

An array of Macintosh workstations running Photoshop, Illustrator, and Freehand, as well as a suite of public domain NCSA software, are a part of this garden of architectures. Macintoshes in both the Numerical Laboratory and the School of Art and Design were used to create texture maps for the Wavefront animations. Images and data were shipped between workstations over networks using NCSA's Telnet software.

NCSA's Visualization Program has provided support software for the REL. Originally, this software was developed for animation production at NCSA (e.g., software to convert files and control animation jobs). The Visualization Program has provided instruction and consulting on custom software, Unix, and production. NCSA's media services have provided consulting and post production for storing, titling, and editing of finished computer animation frames. Critical technical support and systems administration have been provided by NCSA's operations group.

## The Class Structure

The Experimental Visual Technologies (Exp Vis Tech) course has been taught several years and the structure of the class has evolved to consist of approximately 60 percent artists and designers of both technical and non-technical training, 30 percent computer science and engineering students, and 10 percent music composition and other skills. There is a high attrition rate among students in the class during the first semester due to the overwhelming technical information that they must master. The second semester involves a major collaborative effort. Communication skills among students are emphasized. Students evaluate each others' collaborative performance and communication skills. Students are expected to have mastered most of the basic technology including the Unix operating system, vi editor, and a minimal amount of C-shell scripting; three-dimensional animation software such as Wavefront Technologies; and basic storyboarding. Students are also required to be familiar with paint programs such as Photoshop or Targa Tips. A good portion of Exp Vis Tech is concerned with these training issues. In addition, the teaching of NCSA's homegrown software requires a significant amount of time.

The second semester involves the creative application of the technology. A collaborative project is outlined by the professor. Communication skills among students are emphasized. Students are expected to have mastered most of the basic technology. Advanced imaging and animation techniques are coupled to specific project goals. The major project storyboard is assigned in segments to students during the second semester. The ability to follow a storyboard is a skill that is emphasized during the final project.

A large amount of training to prepare a student for this class does not necessarily require computers. Storyboarding and conceptualization are absolutely imperative, yet these skills do not require computers, only ideas, pencil, and paper.

The professor evaluates students and their respective talents during the first semester, then assigns specific tasks on the collaborative project in the second semester. Students are initially assigned to segments according to their strengths and interests. These functions often shift as the project gets underway. The professor attempts to allow students the greatest flexibility. Every student gets introduced to every aspect of the production; however, most students tend naturally to specialize. The production is large enough to preclude most students from grasping the entire scope of the project during the beginning part of the collaboration. One of the most difficult aspects of teaching in this mode is to provide the overall picture to students while they are concentrating on segments. Only near the end of the production do students begin to see the timing and overall complexity of the final product.

The television, computer graphics, and film industries have provided models for the various functions involved in production. There are modelers, animators, texture painters, matt painters, technical directors, art directors, gaffers, production assistants, computer systems support, and camera people as well as other professions in the business. Students are assigned to several computer graphics categories according to their strengths that had been evaluated in the first semester. Highly motivated and proficient students tend to master several areas. It is rare that a student masters every aspect of the production.

Film, advanced computer graphics animation, and television productions are collaborative endeavors. The Exp Vis Tech class is organized after these types of productions and has provided an educational mechanism to teach collaboration as a major component in computer animation and scientific visualization.

## Project 1: Storyboard "Venus & Milo"

"Venus & Milo" is a computer animation narrative set in a synthetic art museum. The museum is a pastiche of old and futuristic artifacts, real and simulated; and each artifact relates in some way to the history of art, science, mathematics and technology. The narrative is centered around the Romboy Homotopy, a mathematical topological deformation (Francis [21], Cox [11]). The initial collaboration to visualize this homotopy was called the Venus Project and has served as a prototype for the Renaissance Team. Frames from this project are shown in Color Plates 33–36.

The Venus is a mathematical surface. The name has evolved from the hour-glass female shape of this surface. She is one step in the evolution of a mathematical transformation. During Venus & Milo, the transformation is the scientific visualization portion of the animation. "For eyes only, the Etruscan Venus ... exists only in the untouchable world of computer graphics... the Venus and her companions are actually visualizations of complex equations created by a mathematician, an artist, and a programmer working together" (Ward [42]). Venus is a phantasmagoric character in Venus & Milo, one of two main characters. The creation of the mathematical transformation is similar to metamorphic imagery that has been explored by artists for centuries. In particular, these works are inextricably linked to the history of caricature and female sensuality (Varnedoe [40]). Milo is a semi-transparent janitor of this unusual art museum. A complete analysis of Venus & Milo, the symbols and art and science history as it relates to postmodern culture is to be published in (Cox [16]).

## *Project 2: Storyboard "Garbage"*

The "Garbage" storyboard was inspired by a major environmental crisis facing our country. The Environmental Protection Agency (EPA), Raleigh, North Carolina, provided some of the statistics for content. The primary goal of this project was to visually and aurally communicate the impending solid waste crisis in the United States. The computer animation was intended to be creative and entertaining as well as have an important message.

"Garbage" begins with a common, metal garbage container looming over the viewer. The title, "Garbage," has been spraypainted on the side of the can. The lid blows off and the can bulges and erupts with litter: plastic and paper containers, used items, aluminum cans, beer bottles, and the like (Color Plate 37). An aluminum can fills our view; a cut to the next scene reveals this refuse dropping to the bottom of the garbage resting on the cement. Finally a used toothpaste tube drops and sinuously comes to life (Color Plate 38). In an extreme close up we see its cap pop open, and toothpaste oozes out of the tube while a man's voice states an innocuous statistic: "Everyday Americans use enough toothpaste to encircle the globe." The tube squirts a path of red-,white-, and blue-striped toothpaste, leading the viewer down the street, leaving a trail through more garbage: pizza boxes, tin cans, plastic bottles, styrofoam cups, and plastic razors. During this trek, the audio becomes dense with female and male voices stating alarming facts about the amount of solid waste that Americans produce: in one week, enough cans and bottles to wrap the globe four times; in one year, two billion razors, two billion batteries, fourteen million tons of cardboard boxes, twenty-six World Trade Centers filled with glass, and on and on. The camera continues to wind through debris until finally see an old newspaper's headline stating "Landfill Close to Full" (Color Plate 39). In the next scene, the camera follows a city street where garbage cans, plastic bags, and houses are popping up, as trees are being sucked into the ground. All of these events have synchronized sound. Five hundred thousand trees are cut down for newspapers every Sunday is a little-known fact that is stated. More facts continue to echo. The view is dense with motion and the proliferation of stuff popping up seemingly is out of control (Color Plate 40). The audio becomes dense with statistics to the point of becoming noise pollution itself. The camera continues to pull back revealing more landscape. The camera cross-dissolves to the next scene of the earth, matching the form of the two horizons as a voice warns us that in five years we will have used over one-third of all landfills. A glass arrow emerges from the earth's cloudy atmosphere. As it moves forward, two more arrows, one made of cardboard and and the other made of metal, emerge and form the recycle logo. A child's voice admonishes, "Hey, get it together. Recycle your stuff!"

The entire animation is filled with wry humor. A spray paint can is labeled "BLOZONE." An old toothpaste box is entitled "Bright Byte." Two-liter plastic bottles contain "Generic Diet" soda. Beer is Light Stout. A green paper flyer inviting people to a meeting on recycling blows in the wind. A green aluminum soda can with lemons and limes on the label is called "Pixel," which is a reference to Sprite soda. (Apple Computer named their frame-buffer picture elements "Sprites." Most computer companies use the term "Pixels.") One of the students modeled the popping houses after the author's house. The pizza box has "Milo's Pizza" with a schematic of Milo's face, a reference to last year's project, Venus & Milo. And, of course, the stars in the background include the planet Venus.

## Conclusion

This experiment has been a learning experience for the author as well as her students. It has been the author's experience that it is best to predesign a storyboard for the group project and teach the students how to follow the storyboard. The initial storyboard often is changed to meet certain needs and accommodate computing schedules. The primary goal here is to teach students collaboration and how it works in a creative, production environment. Students' evaluations of each other are enlightening for each of them. This is an important learning experience that transcends a simple grade for the course. Peer pressure provides motivation for being responsible, on-time, and attentive to others.

While the author conceives and writes a basic storyboard, students and NCSA professionals have creative input during the production. For example, most of the above Garbage labels resulted from students' brainstorming. On Garbage, much of the storyboarding and camera moves were created by

Bob Patterson. In Venus & Milo, the ending was created and modified by Chris Landreth. Students have creative input in designing objects and often make suggestions as to what the objects should be. For example, in Venus & Milo many of the art objects were suggested by art students. In addition, many details are massaged along the way, like in the making of a film. A very important goal here is to encourage people to have "ownership" of the work while at the same time dealing with the reality that the final word is given by the producer or director. Both of the projects are available on VHS video tape in the SIGGRAPH '91 video review.

The computer graphics production environment provides an excellent model for teaching collaboration in computer animation and scientific visualization (Cox, [15]). It is the author's belief that higher education should involve the team approach whenever possible. A large-scale, computer graphics animation is an excellent mechanism for this teaching philosophy because of the complexity of the medium.

## Acknowledgements

I would like to sincerely thank those collaborators who have proven to me that it is truly possible to create innovative, meaningful art as a synergistic, creative team process. Thanks to Robin Bargar for excellent sound and music. Thanks to Chris Landreth, Fred Daab, Michael Moore, Chris Waegner, Marc Olano, Jan Moorman, Christian Erickson, Chris Swing, Ray Idaszak, Ellen Sandor, Stephen Meyers, Dan Sandin, Tom DeFanti, Robert Jordan, Renee Conrad, Joel Knocke, Mike Brandys, Gisela Kraus, George Francis, Maxine Brown, and all of the other students and colleagues who have contributed in one way or another to these projects. Special thanks to Bob Patterson who has been particularly valuable in bridging NCSA groups and has been a primary contributor to Garbage. Mike McNeill and Mark Bajuk provided valuable consulting to the REL in the past year. Amy Swanson has provided critical systems administration support and this is truly appreciated. Thanks to Jim Clark, Forrest Basket, Jack Hanna, and Rick Wells for SGI donations that have made the REL possible. Such an experimental course in this complex, technical environment could not succeed without the software and hardware support that has been provided and partially funded by the University of Illinois, Urbana-Champaign, National Science Foundation. And, of course, a very special thanks to Larry Smarr and the National Center for Supercomputing Applications' staff for continual encouragement and support for creative education and research.

## Glossary

*Postmodern* — an art and social criticism term that originated in fine art photography and that describes art following modernism, developing during mid-twentieth century. Postmodern criticism reveals a hypersensitive historical awareness and recognizes the print and electronic media as central to the transmission of culture and suggests that electronic media has created an isolation of the individual from direct experience of reality; critics observe that art from postmodern era involves a pastiche of recycled styles.

*Renaissance Team* — a group of specialists with complementary areas of expertise who interact synergetically by pooling their technological, analytical, and artistic abilities to increase the domain of available problem-solving options. For example, teams of artists and scientists in the fifteenth and sixteenth centuries produced classic advances in botany and anatomy; their published works are milestones in the history of science.

## References

[1] Baker, W., *NASA: America in Space*. New York: Michael Friedman Publishing Group, 1986, p. 98–111.
[2] Brown, J. and S. Cunningham, *Programming the User Interface: Principles and Examples*. New York: Wiley, 1989.
[3] Brown, J. et al, eds., *Computer Graphics Career Handbook. Computer Graphics*, 23:1, (February 1989).
[4] Burns, S., J. Cohen, and E. Kuznetsov, "Multiple Metamers: Preserving Color Matches under Diverse Illuminants," *Color Research and Application*, 14:1, (February 1989).
[5] Burson, N., "Composite News," *SIGGRAPH Video Review* 11 (1983).
[6] Burson, N., R. Carling, and D. Kramlich, *Composites: Computer Generated Portraits*. New York: Beach Tree Books, 1986.
[7] Careri, Giogio, "Art and Science in Search of Non-Visible Worlds," *Leonardo* 19:4 (1986): 275.
[8] Cohen, J. B., "Color and Color Mixture: Scalar and Vector Fundamentals," *Color Research and Application*, Volume 13, Number 1, (February 1988).
[9] Cox, D. J., "Beyond the Traditional Approach," in J. Brown, ed., *Educator's Workshop Course Notes*. SIGGRAPH '87 (July 27–31, 1987): 207–223.
[10] Cox, Donna J., "Interactive Computer-Assisted RGB Editor (ICARE)," *Proceedings* of the 7th Symposium on Small Computers in the Arts (October 8–11, 1987): 40–45.
[11] Cox, D., "Using the Supercomputer to Visualize Higher Dimensions: An Artist's Contribution to Scientific Visualization," *Leonardo* 21(1988), 233–242.
[12] Cox, D., "The Tao of Postmodernism: Computer Art, Scientific Visualization, and Other Paradoxes," ACM SIGGRAPH '89 Art Show Catalogue, Computer Art in Context, *Leonardo* Supplemental Issue (1989): 7–12
[13] Cox, D., "The Art of Scientific Visualization," *Academic Computing*; March 1990; pages 20–40; references at end of journal.
[14] Cox, D., "Scientific Visualization: Mapping Information," *Proceedings*, Ausgraph '90, Ed. Michael Gigante, September 10–14, 1990, 101–106.

[15] Cox, D., "Scientific Visualization: Collaborating to Predict the Future," *EDUCOM Review*, 25:4 (Winter 1990), p. 38–42.
[16] Cox, D., "Cariacature, Readymades, and Metamorphosis: Visual Mathematics in the Context of Art," *Leonardo* (to appear spring 1992).
[17] Ellson, R. and D. J. Cox, "Visualization of Injection Molding," *Simulation: Journal of the Society for Computer Simulation* 51:5 (1988):184–188.
[18] Fisher, H. T., *Mapping Information: The Graphic Display of Quantitative Information.* Cambridge, Massachusetts: Abt Books, 1982.
[19] Franke, H. W., *Computer Graphics—Computer Art.* Berlin: Springer-Verlag, 1971.
[20] Freidhoff, R. M. and W. Benson, *Visualization: The Second Computer Revolution.* New York: Harry N. Abrams, 1989.
[21] Francis, George, *A Topological Picturebook*, New York: Springer-Verlag, 1987.
[22] Gleick, J., *Chaos, Making a New Science.* New York: Penguin Books, 1987.
[23] Hamilton, G. H., *Painting and Sculpture in Europe 1880-1940.* England: Penguin Books, 1983.
[24] Inose, H. and J. Pierce, *Information Technology and Civilization.* H. Freeman and Company, 1984.
[25] Janson, H. W., *History of Art.* Englewood Cliffs, NJ: Prentice-Hall, 1978.
[26] Kuhn, T., *The Structure of Scientific Revolutions*, 2nd Ed. Chicago: University of Chicago Press, 1962, 1970.
[27] Lucie-Smith, E., *Art in the Seventies.* Ithaca, NY: Cornell University Press, 1980.
[28] McCormick, B.; T. A. DeFanti, and M. D. Brown, *Visualization in Scientific Computing. Computer Graphics*, 21:6 (November 1987).
[29] Meyer, G. and D. Greenberg, "Perceptual Color Spaces for Computer Graphics," *Computer Graphics* 14:3, SIGGRAPH '80 Conference Proceedings, July 14–18, 1980.
[30] Naisbitt, J., P. Aburdene, D. W. Onstad, J. V. Maddox, D. J. Cox, and E. A. Kornkven, "Spatial and Temporal Dynamics of Animals and the Host-Density Threshold in Epizootiology," *Journal of Invertebrate Pathology* (January 1990).
[31] Regis, Ed, *Great Mambo Chicken and the Transhuman Condition: Science Slightly Over the Edge.* New York: Addison-Wesley, 1990.
[32] Rosler, M., "Lookers, Buyers, Dealers, and Makers: Thoughts on Audience," in M. Tucker, ed., *Art After Modernism.* Boston: David R. Godine, 1984, pp. 311–339.
[33] Robertson, P.R. and J.F. O'Callaghan, "The Generation of Color Sequences for Univariate and Bivariate Mapping," *IEEE Computer Graphics and Applications* 6:2 (February 1986).
[34] Ronan, C. A., *Science, Its History and Development Among the World's Cultures.* New York: Hamlyn Publishing Group Ltd, 1982.
[35] Segard, M., "Artists Team Up for the Future," *New Art Examiner* 11:4, (January 1984), pp. 1, 26.
[36] Smarr, L., "An Approach to Complexity: Numerical Computations," *Science* 228:4698 (April 26, 1985): 403–408.
[37] Smarr, L., "The Computational Science Revolution: Technology, Methodology, and Sociology," in Wilhelmson, R. B., ed., *High-Speed Computing: Scientific Applications and Algorithm Design.* University of Illinois Press, 1987.
[38] Sobchack, Vivian, *Screening Space: The American Science Fiction Film.* New York: Ungar, 1987.

[39] Tufte, E. R., *The Visual Display of Quantitative Information.* Princeton University: Graphics Press, 1983.
[40] Varnedoe, K. and A. Gopnik, in *High & Low: Modern Art, Popular Culture.* ed. by J. Leggio, New York: Museum of Modern Art, 1991.
[41] Wallis, Brian, ed., *Art After Modernism: Rethinking Representation.* New York: New Museum of Contemporary Art; Boston: David R. Godine, 1984.
[42] Ward, F., "Images for the Computer Age," *National Geographic*, 175:6 (June 1989), pp. 720–721.
[43] Wead, G. and G. Ellis, *Film: Form and Function.* Boston: Houghton Mifflin, 1981.
[44] Weininger, S., "Science and 'The Humanities' Are Wedded, Not Divorced," *The Scientist*, January 8, 1990, pp. 15, 17.

# Portability of Educational Materials Using Graphics

### Bernard Levrat

Education must benefit from the vast promises of the introduction of computers in schools. The interrelations between the scientific, engineering, and commercial aspects of informatics and the needs of education must be understood, and appropriate measures should be taken to make better use of the intellectual and material resources already available today.

There is no hope of insulating the educational system from the fast pace of innovation. The very existence of computer systems depends on their success in the marketplace, determined mostly by business users. The quest for a really inexpensive machine dedicated to education has not yet produced convincing results in spite of efforts in England, France, and Norway among others. When manufacturers propose somewhat downgraded systems to be sold more cheaply to schools without undermining commercial sales, they generate frustrations among the teaching staff because some of the most attractive software won't run on these systems. Developers are also frustrated by having to contend with poorer capabilities, especially in graphics, than commercial systems offer.

Yet, after a decade of wild development which ended up with the domination of IBM-compatible PCs running MS/DOS, of Macintoshes and their proprietary features, and of workstations running UNIX, the keywords of change in industry are not only more power, more memory, and better graphics but also portability and protection of past investments.[1] New systems must accommodate existing applications with no need to retrain users: the look and feel or the graphical user interface (GUI) must remain the same.

Does this mean that the coming decade will see some stability through standardization? This is promised for new systems but not current ones. Battles are still raging between different GUIs like the Presentation Manager for OS/2 and Open Look and Motif for future UNIX systems; their common feature is a total incompatibility with software currently running on a PC or a Macintosh.

## Chances for "Hyperbooks" to be a Commercial Success

Books are well known to publishers, educators, and students. They present information with good random access facilities, indexes, and a logical structure which is quite apparent and can be followed if one wishes to do so. The text presentation is often enhanced with pictures and graphics. Good books are kept as references for many years.

One could use the term hyperbooks for educational material available through a computer with the added benefits of interactivity and graphics. In the present state of technology, they run on a particular hardware platform that rapidly becomes obsolete and will disappear either because of poor performance or lack of maintenance. Although the delivery machine need not be a stand-alone personal workstation, there are very few examples of networked solutions with heterogeneous delivery platforms.

So far, publishers and manufacturers have not invested much in educational software with the exception of the PLATO system, which CDC took over from the University of Illinois. Manufacturers, when they are interested at all, produce courseware that runs on proprietary hardware. Portability problems make publishers wary of investing in courseware production teams vastly more costly than those needed to produce a book, where the essential part is done by the authors: scholarly books are prepared in connection with teaching or research, with publishers picking up the end product.

Software companies make money in the educational business. They sell language compilers, authoring languages, programming environments, and packages ranging from word processors to expert systems. They have yet to launch significant programs of courseware development. Is educational software, which has seldom been a source of profit for a company, the victim of a fiercely competitive world where there are so many possibilities that only the most rewarding ones are pursued? Is the rate of technological change, imposed by a majority of users more demanding than educators, such that the educational system is not equipped to follow it? What are the contributions of academic research teams?

## Great Ideas, Prototypes and Demonstrations

Doing research is fun. It is very stimulating using an advanced development system to test ideas coming straight from research in artificial intelligence and to think of entirely new courseware material. For good measure, innovative interfaces involving multiple windows and multi-media communications can be thrown in. Whatever comes out will make an excellent show to promote the good work done at the lab, support doctoral dissertations, and give rise to publications.

It is much less popular to try to adapt the same product to hardware which is readily available in a classroom. With a few remarkable exceptions, little or no effort has been made in this direction, and there is no body of research that will give the pre-conditions and the steps necessary to adapt existing courseware to a different environment successfully. Most often, the running program is taken as a specification, and the software is rewritten for the new environment.

Writing courseware without the benefit of an advanced environment is not nearly as much fun. After the stimulating design phase, the humdrum of writing code on the delivery system itself leads, in the absence of the quality control which is characteristic of industrial production, to many shortcuts. Besides, there is very little recognition to be gained in producing excellent material. Unlike a book, a program is often copied without being quoted and, even when it is protected by a copyright, its authors usually keep their anonymity. In this context, the term "product" should be used instead of "program," indicating that a piece of software should not be presented by its author(s) or given any consideration by reviewers until it reaches the stage of being usable by others, which means having an installation guide and a user's manual.

There are no easy solutions to the problem of producing and disseminating high-quality courseware material. It seems reasonable to favor programs that can be adapted to different environments while preserving their intrinsic qualities. To allow for linguistic and cultural differences, these modifications must be made locally. This ability to be locally modifiable without losing its qualities is also a condition of acceptability by local teachers. They will only feel as comfortable with computers as they are with books when they can tailor the the programs' use to their own views and not be forced to use strategies designed elsewhere without their being consulted.

The early success stories of the application of computers are to be found in a number of scientific fields, where the processing power and graphics capabilities of the machines combine with well-designed man-machine interfaces to give users fantastic problem solving capabilities. The most striking examples are to be found in astronomy, crystallography, chemistry, earth sciences, physics, and medicine as well as in many disciplines of engineering, including the more commercial computer-aided design (CAD) systems.

There are lessons to be learned from the way these applications developed. The pioneers understood their discipline very well and also understood the possibilities of applying computers to their problems. After initial demonstrations, the projects received appropriate funding with strong support from their scientific community. It took several development cycles to reach the degree of sophistication observed today, but a conscious effort was made to ensure some degree of portability. When the products reached the marketplace, often through companies with strong links to university laboratories, the psychological problems related to the man-machine interface had been solved, the usefulness of the product had been demonstrated, and the whole scientific community of that discipline was aware of its existence. These remarks are not confined to the hard sciences and engineering. There are beautiful examples in geography, sociology, history and the humanities, like the *Thesaurus Linguae Graecae,* a compilation of all that is known about ancient Greek texts, widely distributed among scholars. It is an ongoing work which, among other things, produces and updates a CD ROM version containing forty-two million words [2].

## *Delivery Platforms*

One way to ensure transferability of applications between systems would be to force any publicly funded project to adhere to accepted standards and not buy non-standard systems. This looks fine on paper but is not practical, because the world of standards is sometimes as strange as Alice's Wonderland.

A standard is usable only if there are available products that implement it. Unless one starts from scratch, these products must also offer a degree of compatibility with what is already in use and for which a large investment has been made. The problems of standards are compounded by *de facto* industry standards. The best known example is the class of IBM PC-compatible machines, which encompasses the IBM product itself and a large number of clones which run the same programs with varying degrees of success. Sophisticated users got good bargains, but many people were caught in a game where there are no written rules. The worst happens when there is hardware dependence, especially when buying a device which is to be plugged directly to the bus, like so many extension cards for graphics, scanners, or network connections.

A PC or a workstation is an industrial product. The "standards" around it tend to protect the interests of its manufacturers. What international conventions can define are interfaces, communication protocols, operating systems, and programming languages. It takes a long time for standards to get all the necessary approvals: the X3J3 committee has been struggling for the last ten years towards a new FORTRAN, not yet available. These delays are explained by the many vested interests in the definition of a new standard.

It is sometimes argued that adhering to standards in a field that is changing so rapidly may stifle research efforts to break new ground. This may be true, but it should never be used as an excuse for not adhering to existing standards whenever possible. As users of informatics systems, members of the educational community should favor standards when they are available and try to create its own *de facto* ones when necessary.

The years to come could see less dominance by a few microprocessors as more and more manufacturers learn to make their own Reduced Instruction Set Computers (RISC). If mass production and its related economy of scale does not stifle the trend, entirely new functions may be devised while maintaining a kernel which accommodates already developed applications.

In this light, the NeXT computer could be considered an example of things to come. NeXT is a new commercial product introduced as "the machine of the nineties" by Steve Jobs, co-inventor of the Apple II and champion of the Macintosh until he left Apple to start this new venture.[1] The editors of BYTE Magazine give the following description: "The NeXT hardware represents an impressive step forward in areas such as digital signal processing, optical disk storage and VLSI technology. The NeXT system software is a step forward for software technology. The system offers an easy-to-use graphical interface to UNIX and an object-oriented programming environment for programmers and software developers" [2]. The NeXT computer has multimedia capabilities built in. It can play compact-disk-quality stereo sound, generate synthetic sound, or work on sounds or voices recorded through a built-in microphone. The Workspace Manager, which handles the graphical user interface, allows an excellent rendering of black-and-white pictures. The machine has been designed for higher education and may be the platform of the future for a friendly companion for students. The success of NeXT depends on whether of not software developers will support it by providing enough products to lure customers to buy it in spite of its stiff current price tag. The question is: do we wait until everyone has a NeXT or a NeXT-clone available, or do we try to adapt what we develop on our own advanced platforms to existing machines?

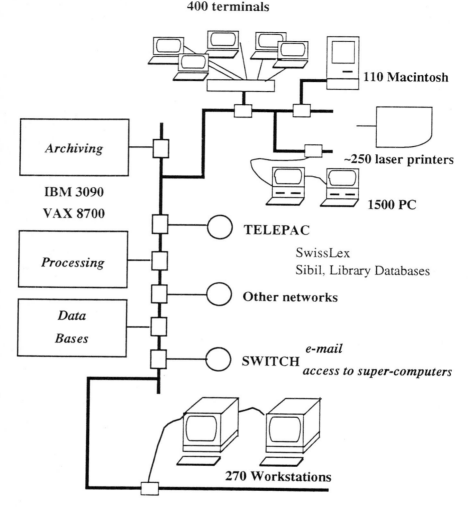

**Figure 1: Abstract view of the University of Geneva network**

## *Immediate Need for Graphics Supported Material on University LANs*

Local area networks (LANs) are becoming common place in universities. They give access to electronic mail, file transfer, and computation on powerful mainframes. Take, for example, the University of Geneva network shown in Figure 1. Any campus-wide application would require interfaces with IBM-compatible PCs made by some twenty manufacturers with a wide range of graphics cards, two generations of Macintosh computers, and all sorts of UNIX workstations. Among the new services we are currently trying either to develop or to generalize on the network, one can mention:

| | |
|---|---|
| campus info | - directories (phone and e-mail), current events |
| telemenu | - access to remote databases |
| friendly mailer | - a self-learning, uniform introduction to any mail system |
| xdico | - on-line dictionaries for six pairs of languages under the X Window System and a full dictionary in French and-/or English |
| encyclopedia | - on-line encyclopedia, with graphical illustrations |
| multi-media | - bring to the desk-top CD-ROMs, video, and sound. |

The first two entries work on all terminals, while the fourth is, at present, restricted to workstations running the X Window System since the current presentation needs full graphics capabilities. It can be seen that managing a campus-wide network brings up the same problems of adaptation as portability of computer-based learning material does.

## *Downgrading Powerful Graphics*

Assume that a wonderful demonstration has been designed on an advanced workstation and that one wants to make it available on the platforms generally available in high schools and universities. Typically, one has to go from a UNIX environment to MS/DOS or the Macintosh operating system. Programs won't transfer easily, even when they are written in a high-level language. Object-oriented approaches aim at the reusability of pieces of software in the same environment. Pascal under UNIX is not comparable to what is offered by Turbo Pascal[1], and a complete graphics package has to be rewritten. C and Modula-2 are slightly better but not without problems.

The use of standards like GKS or PHIGS could provide a solution. They are generally used with FORTRAN or C and are not very popular with people writing educational software. One of the criteria of acceptability of learning material is that it can be adapted locally. This means adapting the interactive dialogs, modifying the menus, and making small adjustments to the presentation of graphics. Doing such adaptations using standards would require a level of expertise rarely found among current users, who prefer to redevelop their own programs and accept a much lower quality than their models but have the fun of doing it all by themselves.

Authoring languages deal more with the logic of the presentation of the material and with the analysis of student's answers than with the design of graphics and their portability to different platforms. If authoring systems with high-quality, portable graphics exist, they have not been given much publicity.

Moving from a high-resolution picture on a workstation to a PC screen probably requires a redesign of the picture. The tools for doing it without too much trouble, using advanced techniques including artificial intelligence, have yet to be developed.

## *Moving Up Towards Higher Resolution*

Some excellent materials exist on PCs, often with the old CGA (200 by 300) standard. When this is moved to a high-resolution workstation it does not look good at all, and it has to be done all over again, which takes time and is utterly uninteresting. Instead, junior programmers assigned to the task play with the program, make it more complicated, and lose a good deal of the pedagogical quality of the original. Here the challenge is to revise the screen design and to improve the pictures in such a way that they do not discourage the learner who is used to other products giving much more stimulating presentations on the same workstation.

## *From Black and White to Color*

It is easy to be convinced that grey-scale rendering of a beautifully colored picture does not always do justice to the creators of the picture. Sometimes it even becomes indecipherable. When software has not been designed with the specific goal of working with the two systems, like Microsoft Windows[1] for example, some rewriting is inevitable.

The main problems when going from black and white to color are concerned with proper choices for lines and background, for contrasting different objects, and for handling correctly the crossing of lines or the superposition of two surfaces. Another problem lies with the attribution of colors among applications, since many overlapping windows can be opened simultaneously.

## *Expected Lifetime of Educational Materials*

Why should we adapt educational materials rather than recreating new ones? The answer lies in the preparation time, which is much longer than an average hardware generation. In five years, what was considered a powerful workstation is outclassed by new products with more MIPS, memory, and storage space than one could have imagined at the beginning. Yet it takes easily five years to design a new course, write the material for it, and test it with students. If one wants a short evaluation by an independent group, it will take a year longer, and one is faced with the problems of scaling up to new hardware. If a hyperbook is successful, people will want it adapted to their current equipment.

By contrast, the book equivalent takes about as long to conceive and to produce. It may go faster for advanced subjects than for basic courses. But once the quality of a book is recognized it has a long lifetime, going through minor revisions as new editions are prepared. Such a book will influence the way a discipline is taught and, in a sense, the way people think about that discipline. If we cannot give our hyperbooks such a lifetime, then we must confine ourselves to preparing laboratory experiments, interesting and rewarding for their authors as well as their users, but leaving most of the curriculum to traditional methods.

One way of extending the useful life of computer-based materials is to save the design in a computer-readable form and to develop automatic program and graphics generation systems. Such a package is part of a research project led by B. Ibrahim at the University of Geneva [3]. In addition to giving a longer life to current efforts, it should allow the reactivation of some of the best materials of the past, if funding can be found and the scripts are still available, even if the supporting hardware has disappeared.

## *The Future of Computers in Education*

Years after the introduction of computers in schools at all levels, they have not made a lot of difference in the way teaching is done. Some brave souls use PCs occasionally to illustrate a lecture or to ask the students to experiment with a model, but by and large, the suggested exercises are drill and practice and most of the classroom time is dedicated to traditional lectures.

My personal feeling is that retraining teachers achieves no more than making them familiar with the computer with all its wonderful possibilities as a text processor, spreadsheet, database, or hypertext support. To include an informatics component in the teaching of their own discipline, teachers have to experience it as students and to see it as the normal way of doing science. They have to own the technology. That is why it is more important to put advanced graphics facilities in universities than in high schools at the present time. Today's students are tomorrow's teachers. Let them define the environment they feel comfortable with to transmit knowledge to others.

Students should also be taught to use computerized material as they are taught to use books. With a consistent interface to a lot of material which survives hardware changes and is readily available, the computer, with its powerful information retrieval and beautiful graphics, will really become the lifetime partner of modern man, eternal student in a dizzyingly fast changing world.

## Notes

1. Unix, MS/DOS, Macintosh, NeXT, Turbo Pascal, and Windows are trademarks of AT&T, Microsoft, Apple, NeXT, Inc., Borland International, and Microsoft, Inc., respectively.

## References

[1] Brunner, T., *Thesaurus Linguae Graecae* Newsletter, University of California Irvine, 1991.
[2] Thomson, T. and N. Baran, "The NeXT Computer," *BYTE*, 13:12 (1988), 158–175.
[3] Ibrahim, B. et al., "Courseware CAD," *WCCE/90 Conference Proceedings*, North-Holland/Elsevier, 1990, 383–389.

# Collaboration between Industry and Academia — Computer Graphics in Design Education

## Adele Newton

Computer-aided industrial design (CAID) is a new and emerging application of computer graphics. Designers of such diverse items as cars, jewelry, furniture, luggage, and toys use computer graphics to design, visualize, and market their products. But are design schools preparing their graduating students for today's design job market? Schools and industry can work together to incorporate computer graphics into design-school curricula. This paper will discuss the benefits of collaboration between design educators and the computer graphics industry and will describe the successful university program at Alias Research as a model of this type of collaboration.

### Why Computer Graphics in Design Education?

Computers have been described as cold, dehumanizing and, by their very nature, counter-intuitive to the creative process. The administrations of some art and design schools believe that the use of computers as design tools will negatively influence designers. These schools have been slow to respond to the challenge of expanding from traditional training methods. To quote Paul Brown, "The basic assumption is that the discipline is secure and that it will not be changed by new technology" (Brown, [1]). Computers as design tools are a fact of life, and industry will continue to demand employees who are familiar with a full range of tools, including computer graphics. Del Coates, Professor of Industrial Design at San Jose State University, predicts that "the computer will be the basic industrial design medium within five years" (Coates [2]).

As recently as ten years ago, computer graphics systems were developed and used mainly by computer scientists on mainframe computers.

Today's advanced three-dimensional workstations and interactive software have revolutionized computing and put the power of computer graphics into the hands of designers. The workstations of the 1990s allow designers to work in an interactive three-dimensional environment where they can sketch, draw, visualize, build, and animate prototype models.

In the current competitive market, the challenge to industrial designers is to shorten the product-design cycle and to respond faster to market trends. Traditional design methods require constant modifications to sketches and renderings as well as the the construction of prototype models of the design. The use of computer graphics allows designers to refine and change their designs easily at any time in the design cycle. Shapes, positions, material properties, and colors can be changed quickly, precisely, and in real time. Surface finishes and lighting are controlled by the designer to produce photographic quality imagery which can replace physical models. These electronic mockups communicate the designer's vision faster and better to clients, engineering departments, and upper management.

## *How Technology is Changing Design Education*

The first step in the process of integrating technology into design education is for educators to recognize that they have nothing to fear from technology. Computers cannot replace designers but can serve as communication and presentation tools. The schools that train tomorrow's designers cannot remain isolated from the technology used by the companies that will hire their graduates.

Companies such as General Motors, Honda, Timex, Oneida, and hundreds of others worldwide that use CAID software are sending the message loud and clear: the use of computer graphics in the design process is absolutely critical in today's competitive global marketplace. Design educators who ignore that message do so at their peril and at the peril of their students.

The effective integration of technology into design-school curricula will require ongoing collaboration between academia and industry. Design educators must do more than invest in computer hardware and software. They must also take the time to learn what best serves industry and will therefore best serve their students. To stay on the leading edge of technology, schools must maintain an ongoing dialogue with the companies that will employ their graduates and with the suppliers of technology.

## Collaboration between Design Schools and Industry

There is an established history of collaboration between industry and academia. Computing companies such as IBM, Xerox, and Digital Equipment Corporation are well aware of the benefits, both intellectual and practical, of establishing working relationships with universities. These collaborations bring together superior minds from industry and academia and, as a result, all parties benefit.

It is imperative that schools and industry maintain ongoing relationships after the installation of software and/or hardware. A key aspect of those relationships is the integration of software in the graduate or undergraduate curriculum. Students are afforded the opportunity to work on real-world problems. Technology resulting from student and faculty research is applied by industry in engineering and production. Industry is assured of a source of properly trained employees and students of jobs after graduation. Design educators have much to gain from developing and maintaining ongoing relationships with the computing companies that supply computer graphics hardware and software.

## The Benefits of Collaboration

### INTERNATIONAL EXPOSURE FOR STUDENT WORK
Students can ensure international exposure for their work by providing images for use in industry public relations and marketing campaigns. Because students are free to experiment with the features of CAID software and are not bound by the same deadlines and restrictions as commercial customers, they are often able to provide exciting and innovative images for use in advertising, demo reels, and brochures.

### JOBS FOR GRADUATING STUDENTS
Purchasers of CAID software are interested in hiring the best new designers who are proficient users of the software. To meet those needs, software and hardware vendors can provide career placement services for graduates of schools with which they collaborate.

### PRODUCT DESIGN INPUT
While learning to use computer graphics as part of their design education, students can contribute to the design of the CAID tools they will use in the future.

## Features of Successful Collaboration Programs

*THE PROGRAM REQUIRES COMMITMENT FROM BOTH SIDES.*
Schools and industry must continue to work together after the installation of software and hardware. To ensure the on-going success of such collaborations, both sides must make commitments to maintaining the collaboration.

*THE NEEDS OF ALL PARTIES MUST BE MET.*
Programs must be designed to benefit design schools and their students as well as the collaborating industrial partners.

*THE PROGRAM MUST BE FLEXIBLE.*
To keep such programs vital, schools and industry should explore ways to collaborate on new and innovative projects. These can vary from joint research in areas of mutual interest to specific, classroom design projects.

*THE PROGRAM MUST BE RESPONSIVE.*
To ensure that such collaborations thrive, industry and academia must continue to look for ways to serve design educators, students, and industry.

## Other Issues

While forging strong relationships, design educators and industry must address the intrinsic conflicts between the mandate and needs of schools and those of industry partners. Educators and researchers in all fields, including design, are concerned about the prospect of compromising the integrity of their work by becoming too closely associated with industry. This concern is tempered by the need for industry support of education in an era of government underfunding. Researchers and educators in other disciplines have come to realize that collaboration is effective when the program is carefully designed to meet the needs of both sides. Design educators will find that if steps are taken to ensure that this happens, working with industry will enhance and strengthen design education.

Industry is concerned with keeping innovations secret until they have been legally protected as intellectual property and delivered to market. The focus is usually on delivering a completed product, in-spec and on-time. While companies have realized that there is a competitive edge to be gained by collaborating with educators to share access to ideas and people, time must be invested to carefully design programs that address concerns about proprietary information while allowing industry to benefit from collaboration with schools.

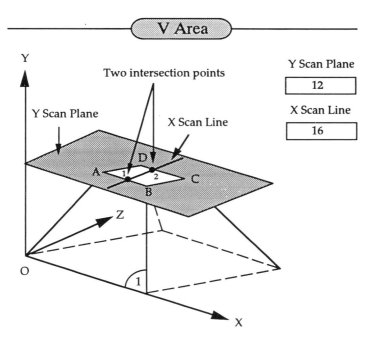

**Figure 1: A scene from the V area of the simulator showing a Y-scan plane and an X-scan line. Two line segments AB and CD intersect the X-scan line resulting in two points. Clearly point 1 is considered as visible and thus face 1 will provide colour information for the related pixel**

During processing, the scan plane moves down the screen from top to bottom. For each scan plane, an active edge list is built, into which are inserted all edges intersecting the plane. As the scan plane moves down the screen, edges which no longer intersect it are deleted from the active edge list. The projection of an edge onto the X-Y plane forms a line segment; projecting each active edge in this way forms a list of segments, which is then sorted in order of increasing minimum X. Next, a scanning process is applied in the X-direction. At each step in this process, an X scan line is intersected with the list of segments, as shown in Figure 1, and a list is built of those segments which span the current X scan line. The minimum or maximum depth value is selected according to the screen coordinate system used. In this paper, the Z-axis is considered to be pointing towards the screen, and thus the minimum value will determine the visible face — that is, the face nearest to the view plane.

## The Simulator

The simulator is an implementation of the algorithm described above for visualization and teaching purposes. Its function is to make it easier for students to observe and understand what is happening during the generation of a real, three-dimensional picture. This section gives a detailed description of the design of the simulator, based on results from a prototype described in the discussion on interactive teaching using the simulator, below.

A popular method in computer graphics applications, especially in computer-aided design and engineering, is the representation of objects by polyhedral networks. Basically, such a network is a set of vertices, edges and faces in three dimensional space. For simplicity, faces are considered here to be planar and the simulator can only handle such representations.

To begin with, the simulator displays a menu of choices amongst which are the creation of objects, the definition of a viewpoint, and positioning of a light source. Dialogue and colour menus are similar to those in a Macintosh interface. Menus appear instantly on the screen and disappear after selection. In the "Operation" menu, for instance, students have the choice of performing shading, visualization, or both. One way of achieving this is to divide the screen into two main areas: a visualization area (V area) and a working area (W area). The latter shows the part of the picture generated so far. In the V area, however, students can inspect what is happening in the W area line-by-line and pixel-by-pixel.

The heart of the simulator lies in the method of visualizing the two scanning processes. Initially, a wireframe picture of the object is displayed in the V area after setting up a particular visualization view environment. To avoid confusion, a decision needs to be made whether or not to display the hidden lines in the scene. Both cases are supported and the user's preference can be indicated via the "View Setting" menu. However, it is better not to display hidden lines for an inexperienced student. Hidden lines can only be displayed while visualizing the building and maintenance of data structures as explained later. Finally, the idea of displaying lines with various intensities according to their visibility status might well be considered as an alternative — for example, by displaying visible lines brighter than hidden ones.

With the Z-axis normal to the screen, it is difficult to visualize the performance of the algorithm. Accordingly, the whole scene can be rotated about the X-, Y-, and Z-axes. This is crucial in making the visualization process clear enough for students to observe the line segments generated by the intersection of the faces with a Y-scan plane. A scene from the V area is shown in Figure 1 which illustrates this point. Furthermore, rotation not only makes it easier to visualize depth information but also allows the user to foresee the visible face. However, it should be mentioned that the rotation is only applied to the picture in the V area. The Y scan plane appears as a parallelogram corresponding to a Y scan line in the W area. Before simulation begins, the user is prompted for rotation, translation, or any other transformations via the "View Setting" menu. The shading environment, comprising details such as colour intensities, light source positions, and other information, is also specified before simulation commences.

## *Visualization of the Y Scan Process*

For each Y scan line in the W area the corresponding Y scan plane is created and transformed according to the view setup. After that, it is displayed in the V area. The question arises of how to display the corresponding cross section of that plane and the object clearly. One possibility is to make the scan plane transparent. However, this may lead to confusion in complicated scenes. This may be acceptable for skillful users, but for those with less experience emphasis should be put on making the view clear enough to highlight the performance of the algorithm. The intention is that the construction process for the line segments forming the cross section can also be visualized, as can the process of building and maintaining the data structures involved.

In building line segments, the Y scan plane is made transparent and the picture is displayed with hidden and visible lines. Either low intensity or dotted lines can be used to indicate the hidden lines in the picture. The first edge $E_1$ in the list of active edges is drawn in a specified colour called ActiveEdgeColor (see Figure 2). Its intersection point A with the scan plane is displayed in a different colour designated by IntersectionColor. After that, the list is scanned for the counterpart edge $E_2$ on a face common to $E_1$, for example on faces $F_1$ or $F_2$. The edge $E_2$ is also displayed in the same colour as $E_1$, whereas the selected face is shaded in its corresponding colour named ShadeFaceColor. The two intersection points are joined to form the corresponding line segment drawn in a colour of value InactiveSegmentColor. Similarly, other line segments are found until the complete cross section is obtained. After that a new scene is generated by removing hidden lines and making the plane opaque again.

It should be noted that, after a Y scan plane process is terminated, those parts of the picture in that plane are deleted from the V area. The inactive edges remain intact whereas the active edges are updated and redrawn. As a result, everything happening in the next Y scan plane will be clearly visible. Gradually, the picture diminishes in the V area and moves into the W area.

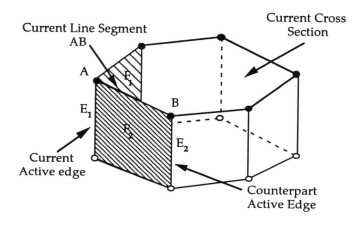

**Figure 2: Visualizing the building of line segments**

## *Visualization of the X-Scan Process*

Each step in the Y scan process results in a number of line segments which are all initially inactive. The first corresponding X scan line is built, transformed, and displayed in the V area using a colour named ScanLineColor. The process begins by highlighting the active line segments that intersect the current scan line and displaying them in a colour of value ActiveSegmentColor. Initially, the intersection point of each of these segments with the scan line is found and displayed on the screen in a colour InvisiblePixelColor. The display of depth information at each of these points might well be considered as an option here.

Before wiping out all invisible points after comparison of depth values, students are given an opportunity to guess which point is visible. The latter will eventually be displayed in a colour denoted by VisiblePixelColor. After computing its colour information, the corresponding pixel in the W area is displayed.

The process of updating the data structure for the line segments can be visualized by gradually shrinking the active line segments and later deleting them completely from the scene. As the X scan line moves towards X = Xmax, the visible active line segments are shown in a colour designated by VisibleSegmentColor. This is achieved by erasing the line segments after each X scan process, updating and redrawing them in their corresponding colours.

## Interactive Teaching Using the Simulator

The simulator not only provides a good environment for students to visualize the performance of the scan line algorithm, but also allows them to gain insight into the problem by interaction. It has two main aspects: a visualization aspect and an interactive teaching aspect. In the former, the simulator will enhance their imagination capability as far as the topological and geometrical information are concerned in related problems. By such visualization, students should learn how faces are classified as either visible or invisible, how the data structures involved, such as the lists of active edges, the list of inactive line segments, and other lists, are built and maintained. Visualization of these lists can be provided, either in the V area or in a separate window.

Skillful users can also benefit from the simulator by directly visualizing the scene at a particular Y scan plane or an X scan line. The desired number of a scan line or scan plane is input and the scene corresponding to this is automatically generated. Furthermore, students can visualize the whole process from different views, with different shading and lighting conditions. In teaching related topics, live demonstrations can also be shown using the simulator, especially when sophisticated scenes are considered. For some students, details of the performance of the algorithm in such scenes are not easy to grasp without visualization.

The system can also be used in an interactive teaching environment, where students are asked to make their own decisions for a particular Y scan plane process in a given scene. Initially, a wire frame picture is displayed and students are asked to construct both the inactive- and the active-edge lists. This can be achieved by clicking on the appropriate edges which are inserted automatically into their corresponding list. One way of visualizing this is by colouring the clicked edges according to the list they belong to. The display of depth and other information by the system will guide the students in making correct decisions.

By coupling edges from the list of active edges, the list of inactive line segments can be built. To begin with, students select (click on) two active edges common to a visible face. After that the corresponding line segment is constructed and displayed. Similarly, other line segments are found until the complete cross section is constructed. Now the X scan process session begins, and students have to indicate the active segments intersecting a particular X scan. After computing and displaying all intersection points, students click on the visible point and later check to see if the correct decision was made. The level of difficulty can be adjusted by controlling the amount of information provided to the students. In this way, the system can be used for intelligent tutoring.

## The Prototype Simulator

A prototype version of the simulator described in this paper has been implemented on a Macintosh IIci, using MPW Pascal. Snapshots of this system in action are shown in Color Plates 43–45, which depicts the process of shading a cube. The active segments are drawn in light blue, and the inactive segment ones are displayed in black. In this example, only two points are generated on each X scan line (Color Plate 43) and these are initially drawn in navy blue. After that, the visible point and the invisible point are displayed in red and dark green respectively.

It should be mentioned that because of the amount of computation resulting from the combination of several processes, the algorithm should ideally be implemented on a more powerful machine.

## Conclusions and Further Work

The full simulator will be developed and used as an aid in teaching computer graphics courses at the United Arab Emirates University and the Lebanese University. In the classroom, live demonstrations will be shown when the topic of hidden surface removal is introduced. Students will interact with the system outside the classroom to fully understand the material involved. Finally, the system itself will be used as a testing tool to ensure that our educational goal is reached. Thus, it is hoped that quality of learning will be improved, allowing for the incorporation of newer materials into related courses.

Armed with these capabilities, such a system is a good candidate for inclusion in a hypermedia system for computer graphics education. For instance, it could be used as a node in the HyperGraph project, sponsored by ACM SIGGRAPH [6]. Or it could be used in a multimedia electronic book on computer graphics, allowing the user to explore the scan line algorithm interactively, rather than just looking at preselected, static output images.

Further work is still in progress on improving the visualization process. For example the display of three scenes simultaneously is being considered: the real scene with shading only, the scene with full information displayed (all lines displayed with a transparent scan plane), and the realistic scene after excluding the hidden lines. Another feature being studied is the capability of simulating at various levels of resolution. This is to enable the generation of the picture either pixel-by-pixel as it happens in the W area, or to enlarge the picture at some critical situations in the V area. Perspective transformation might also beneficially impact the process of visualization.

Finally, the author hopes that this paper will motivate the development of other simulators for non-trivial algorithms or other areas in computer graphics, such as shading, shadows, and texture mapping. He believes that simulations of the type described here will result in a more skillful generation of computer graphics students, paving the way to a better role for computer graphics in education.

## *Acknowledgments*

The author would like to thank Professor Robin Forrest for his encouragement and fruitful suggestions. Thanks are also due to the referees for their valuable comments. Finally, thanks go to the United Arab Emirates University and the Lebanese University for providing financial and moral support.

## *References*

[1] Burger, P. and D. Gillies, *Interactive Computer Graphics: Functional, Procedural and Device-Level Methods.* Addison-Wesley, 1989.

[2] Eckert, R. R., "Kicking Off a Course in Computer Organization and Assembly Language Programming." *ACM SIGCSE Bulletin,* 19(4,) December 1987.

[3] Hays, H. D., "Interactive Graphics: A Tool for Beginning Programming Students in Discovering Solutions to Novel Problems." *ACM SIGCSE Bulletin,* 20(1), February 1988.

[4] Hearn, D. and M. P. Baker, *Computer Graphics,* Prentice-Hall, 1986.

[5] Lane, J. M., L. Carpenter, T. Whitted, and J. F. Blinn, "Scan-Line Methods for Displaying Parametrically Defined Surfaces," *Communications of the ACM,* 23(1), January 1980.
[6] Owen, G. Scott, "HyperGraph — A hypermedia system for computer graphics education," In *Interactive Learning Through Visualization: The Impact of Computer Graphics on Education*, S. Cunningham and R. Hubbold, eds. Springer-Verlag, 1992.
[7] Watt, A., *Fundamentals of Three-Dimensional Computer Graphics,* Addison-Wesley, 1989.

# Solid Modeling in Computer Graphics Education

## Alberto Paoluzzi

The first aim of this paper is to show the influence that graphics methods have had on the research activities carried out in solid modeling at the University "La Sapienza." At the same time, we report our experience about the impact of solid modeling tools, equipped with a user-friendly user interface, on computer graphics education. In particular, we describe the solid modeler *Minerva*, implemented by computer graphics students and currently used as a didactic tool in our graphics course, which allows students to experiment with affine transformations, and to view models, structure definitions, surface equations, and solid operations.

Graphics methods and ideas have also driven our approach to multidimensional solid modeling, a research topic which may give some insight into the possible unification of several problems in CAD, computer graphics, and applied geometry, in the areas of scientific visualization, engineering computation, and robotics. The need for powerful design languages, in order to blend the applicative style of programming with CAD and graphics methods like structures, solid operators, and variational geometry, motivated the PLASM project described in the paper. Such a language should be based, in our view, on a dimension-independent representation of objects and, once again, on interactive graphics methods. To implement effectively such ideas may require huge computing power, but next generation graphics machines are at the door and parallel computers are already here.

## Background

First experiences of computer graphics at the Engineering School were gained with time-sharing graphics terminals and pen plotters at the Computing Facilities Department in the late sixties. A research lab in computer aided building design was created in 1976 at the Institute of Architecture, Building and Town Planning, with a minicomputer, some raster graphics terminals and digitizers and a bed plotter. The work of the lab at that time concerned network programming techniques, the development of abstract models of the design process, and some specialized data structures and data bases.

In 1984 an undergraduate course in Computing Techniques for Architecture and Planning originated from the lab. Two years later the course was transformed into a course in computer graphics for students in Electronic Engineering and Computer Science. The current name of the course is Informatica Grafica, at the fifth year of the new degree in Computer Engineering. It is currently the only offering in computer graphics at the Engineering School. About fifty to seventy students follow the course every year, and about fifteen different projects are developed by groups of two or three students. About five to seven Master theses are also developed in this area every year. The computing equipment mainly consists of Macintoshes. Some workstations with advanced graphics features have been obtained recently.

Currently, the computer graphics students are asked to develop a non-mandatory implementation project, as a partial fulfillment of the requirements for their exam. Every year a different subject is chosen, within the framework of the projects developed by the CAD group, and a set of coordinated tasks is proposed. Every task is then assigned to a group of two or three students. Some of the themes assigned in recent years concerned device-independent graphics kernel and device drivers, hidden surface and rendering algorithms, operations over winged-triangle representations of polyhedra, polyhedral approximation of surfaces.

In our experience, the requirement of developing a complex implementation task within a predefined framework can be considered very positive, both from the student's and from the teacher's point of view. Often some tasks are studied more thoroughly as a thesis topic for the *Laurea* degree in Electronic or Computer Engineering. The best works of students are integrated in our research and development projects.

## Course Programme

The course in graphics and modelling, started six years ago as an introduction to design with computers for undergraduates in Civil Engineering, was then directed to students in Electronic Engineering and Computer Science. The programme of the course changed quite quickly every year. In the first two years, according to the first diffusion of personal computers, the main subject concerned the basic algorithms of computer graphics and the computer graphics standards, with special emphasis on GKS and on the implementation of a simplified, device-independent graphics system. Later the course has developed to include methods of solid modelling, paying special attention to the boundary representation of polyhedral objects and to the definition of solid primitives and operations, like extrusion, revolution, and Boolean set operations. Finally, curved surfaces with parametric equations have been dealt with as well.

The current programme of the course in "Informatica grafica" is subdivided in two nearly equivalent sections, concerning Computer Graphics (twenty-eight hours) and Geometric Modeling and standards for Computer Graphics and CAD (thirty-two hours).

In our experience, as graphics hardware becomes more and more powerful, less attention can be paid to basic algorithms in computer graphics and rendering (clipping, filling, hatching, rasterization, shading). A good understanding of the affine and projective transformations is, conversely, always necessary, together with the mastering of higher-level primitives and operations concerning surfaces and solids.

With respect to this basic educational objective we have found it very useful to use as a didactic tool our solid modeler *Minerva*, described in a following section. The use of this system in the laboratory allows students to increase their geometric perception of space, to experiment with the effects of elementary transformations, to understand the differences between various kinds of projections, and to master the use of local coordinate systems in the definition of structures containing objects and transformations of coordinates. The distribution of wide sections of the *Minerva* code, in most cases linked to the student's personal project, allows the student to perform non-trivial programming within the framework of a large software project.

1. Computer graphics
   - Affine transformations [4 hours] (translation, scaling, rotation, shearing; composition and properties of transformations)
   - Transformation pipeline [6 hours] (modeling, view mapping, perspective, normalization and workstation transformations)
   - Basic algorithms [6 hours] (traversal; line and polygon clipping; point inclusion; hatching, filling, rasterization; z-buffer)

- Advanced algorithms [6 hours] (generalized polygon clipping, depth sort, parallel depth merge, line and area coherence, binary space partition)
- Rendering [6 hours] (taxonomy of projections; camera model; shadow computation; colour models; lighting models, constant, Gouraud and Phong shading)

2. Modeling and standards
   - Curves and surfaces [9 hours] (implicit, parametric and intrinsic equations; revolution surfaces, ruled surfaces, cones and cylinders, Coons surfaces; algebraic and geometric forms of a parametric equation; Hermite, Bezier, cardinal spline, B-spline, NURBS)
   - Standards [6 hours] (GKS, CGI, CGM; GKS-3D, PHIGS, PHIGS PLUS; IGES, STEP)
   - Basic solid modeling [8 hours] (representation scheme, Requicha's taxonomy; space occupancy, object decomposition; CSG; boundary Reps, winged-triangle rep)
   - Advanced polyhedral solid modeling [9 hours] (integration, extrude, revolve, set operations; simplicial complexes, winged rep, simplicial approximation of multi-dimensional manifolds and fields)

## Graphics Interfaces to Modelling

In this section some features of our projects in solid modelling of polyhedra are summarized, paying special attention to the interaction environment they provide. *Minerva* is a polyhedral solid modeller based on a straightforward, multiple-window interaction environment which fully exploits the Macintosh user interface. *Simple* $_X^n$ is a prototype multi-dimensional modeller which is being developed to show that dimension-independent methods may allow us to treat, in a unified manner, several graphics and geometrical problems, including graphics queries, graphics rendering of fields over manifolds (some kind of scientific visualization) and motion planning. *PLASM* is a prototype design language also based on dimension independent methods. We believe that a key point for the language will be a well-designed, visual user interface. Such an interface should provide a language translation of any graphical action of the designer on the currently selected shape (picking, grouping, cutting, pasting, dragging), as well as providing graphical echoing of language expressions evaluated in the listener window. Furthermore, the language should be considered a component of a very high level graphical environment, with solid modelling, rendering, and real-time visual simulation subsystems. Our aim is to provide standard interfaces toward such subsystems, and, at the same time, a complete independence of the language, so that it will be possible to exploit future hardware and software developments. The *Minerva* project started in 1985 and was concluded in 1990. Both the *Simple* $_X^n$ and the *PLASM* projects started in 1989.

## Minerva *Solid Modeler*

*Minerva* is based on a boundary representation called a "winged-triangle," which is a triangular decomposition of the object boundary and results in a very simple data structure, as any boundary triangle has three vertices as well as three adjacent triangles. The representation is very efficient; it is possible to prove [10] that in the best case — when linearizing the boundary of curved solids — the representation needs exactly half the memory space of Baumgart's well-known "winged-edge" representation. *Minerva* supports most of the operations that are performed by solid modelers. In particular it can (a) generate simple primitive solids (a cube, a sphere, a torus); (b) generate extruded and revolved polyhedra starting from polygons; (c) combine simpler solids with Boolean operations (complement, union, intersection, difference); (d) generate structured solids in a PHIGS-like manner; (e) render realistic scenes by removing hidden surfaces and performing colour shading (constant, Gouraud and Phong); (f) compute integrals of polynomials (volume and moments of inertia) over polyhedral solids and surfaces [3]; (g) generate parametric surfaces, via the symbolic definition of their equations [7]. Figures 1–3 introduce the environment and capabilities of Minerva.

Polyhedra and structure files can be stored both as memory image and as text files. Polyhedra can also be exported as DXF files (AutoCAD format) in order to be transported to external renderers. Pictures generated by *Minerva* can be copied in PICT format to the Macintosh clipboard to be imported (pasted) into external drawing programs such as MacDraw or Claris CAD. The importing of externally defined sets of polygons is also allowed. Such polygons can be extruded by *Minerva*, both as a multishell polyhedron and as a structure composed of different solids. Structures can be graphically edited by interacting with three orthographic views. Single

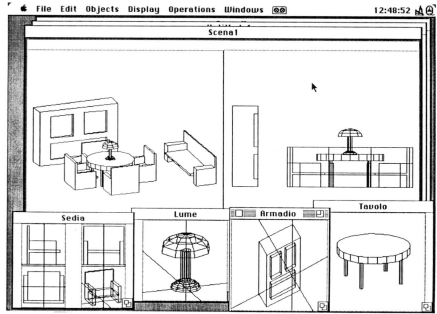

**Figure 1:** The window-based user interface of *Minerva*, with a structure definition

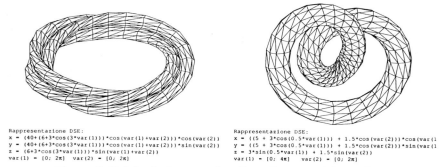

**Figure 2:** Polyhedral approximation of parametric surfaces generated by *Minerva*

components of a structure can be picked, and then scaled, translated, or rotated. Symbolic editing of structure elements is also allowed.

In any object window, *Minerva* can generate one, two, or four projections. Hidden surfaces can be removed by a combination of depth-sort and front-to-back presentation with generalized polygon clipping. Twelve pro-

# Solid Modeling in Computer Graphics Education 233

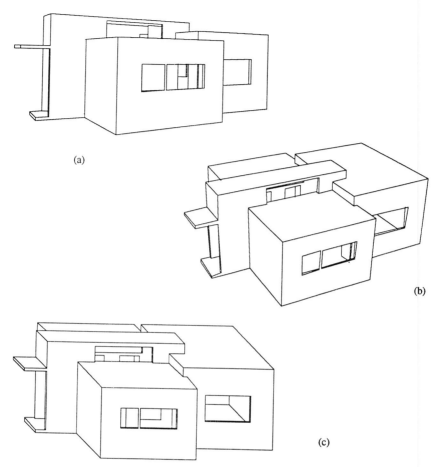

**Figure 3: Solid model of the Oud's Mathenesse (Rotterdam, 1923) generated by *Minerva*. (a) Two-point perspective; (b) three-point perspective; (c) dimetric Cavalier axonometric view**

jections are predefined, including three orthographic views, three orthogonal axonometric views (isometric, dimetric, trimetric) and six oblique axonometric views (various isometric and dimetric cabinet and Cavalier views). Other parallel or projective views are obtainable by asking the user to define the View Reference Point (VRP), the View Up Vector (VUV), the View Plane Normal (VPN) and either the Direction Of Projection or the Center Of Projection (COP). Such a user interface for the definition of projection parameters has been found to be very useful in the context of a graphics course, for teaching the geometry of projections and the "camera model."

In order to link the camera model to perspective theory, in the *Minerva* interface to projections we have chosen to define COP in a coordinate frame parallel to the world coordinate (WC) system and with the origin in VRP, and we have applied the constraint VPN = COP. In this way we obtain a one-point, two-point, or three-point perspective depending on the number of non-zero coordinates in COP. In perspective theory such projections correspond to the central, accidental, and oblique perspective, respectively.

The allowed operations include union, intersection, difference, and complement of polyhedra (see Figure 4). The simple algorithm used, based on the computation of the intersection of each triangle pair, is described in [10]. The Boolean operations are closed over the set of non connected, non bounded (but with bounded boundary), and non-manifold polyhedra. An integration menu allows computation of the external surface, the volume, the centroid, the first moments, the products, and the moments of inertia with respect to the WC axes. Such integrals of a low-degree polynomial over a polyhedral domain are computed by reducing the integration to the boundary of each triangle of the object surface, by iterative application of the Gauss theorem. The integration algorithm is described in [3].

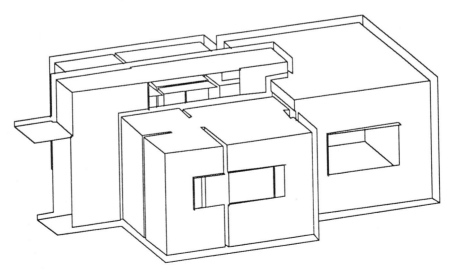

Figure 4: Hidden surface removed view of the complemented Mathenesse building

The solid modeling and rendering operations are built on the top of a GKS-3D-like, device independent graphics system extended with structures. Interaction is controlled via a powerful user-friendly, multiple window, graphical interface currently implemented in a Macintosh environment [9]. *Minerva* can be obtained for educational purposes by contacting the authors.

## Simple $_X^n$ Multidimensional Solid Modeler

*Simple* $_X^n$ is an experimental software system for generating and rendering dimension-independent polyhedra, i.e., polyhedra with intrinsic dimension $n$, possibly embedded in a space $\Re^m$, $1 \le n \le m$. *Simple* $_X^n$ supports both decompositive and boundary representation of linear polyhedra using a simplicial based representation [8] which is a direct extension of the *Minerva* winged-triangle representation.

A multi-dimensional approach makes it possible (i) to unify solid modeling methods (three dimensions), motion encoding and interference detection (four dimensions), and to represent (($d+k$) dimensions) the whole set of configurations of a $d$-dimensional system with $k$ degrees of freedom; (ii) to introduce simple and general methods for representation and rendering of fields, e.g., the evaluation of "geometrical queries" over fields, i.e., the computation of subsets of points which satisfy a predicate calculus formula (where atomic sentences are relational expressions with field variables), is reducible to a Boolean formula with $n$-dimensional manifolds; (iii) to reduce the gap between solid modeling and finite element and multigrid methods. For example, FEM discretizations can be mapped to decomposed solid representations, and the field values can be stored as additional coordinates of vertices.

At the current time *Simple* $_X^n$ allows (a) generation of higher dimensional polyhedra starting from lower dimensional ones, via extrusion and screw-extrusion operations [6]; (b) the evaluation of Boolean expressions between multi-dimensional polyhedra and their reduction to a canonical form [2]; (c) the integration of polynomials over polyhedral domains [1]; (d) the evaluation of structured objects defined by using local modelling coordinates and their reduction to the same world coordinate system by the traversal of their symbolic description; (e) the definition of simplicial maps, which allow generation of polyhedral approximations of curved manifolds and fields over manifolds [12]. See Figure 5 for some examples of these operations. The *Simple* $_X^n$ system is documented in [4]. The system, initially

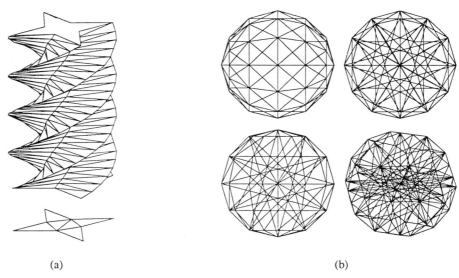

**Figure 5:** Two polyhedral objects generated in a dimension-independent manner: (a) the boundary of the screw-extruded polyhedron generated by a decompositive representation of the two-dimensional star; (b) some three-dimensional projections of a five-dimensional polyhedron

developed in Common Lisp in a Macintosh environment, is currently being transported to workstations working under UNIX.

## PLASM *Language*

*PLASM* is a programming environment for performing a sort of polyhedral calculus over polyhedra. It is a functional language which supports the writing of scripts which are sets of definitions of (parametric) solid objects. In this language a number of new operators are given; using such operators it is possible to define and manipulate easily and compactly a dimension-independent polyhedral complex, defined as a cell complex whose cells are polyhedra. As well as the standard Boolean, extrusion and integration operations given by any solid modeler, the language introduces new powerful operations for product, power, offset, etc., of polyhedra. A slight extension of the traversal operation, typical of PHIGS-like graphical systems, is also given, which allows us to describe structured, solid objects and environments fully without detailing their low-level geometry. Such operators have been invented in the framework of a project for computer aided architectural design, but can also be used for general applications, particu-

# Solid Modeling in Computer Graphics Education

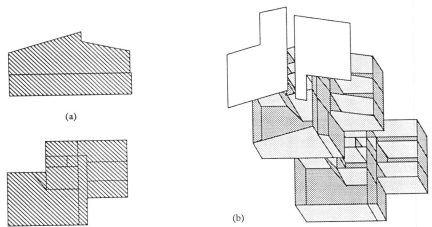

**Figure 6:** (a) The arguments *Plan* and *Section* of an operation && (intersection cell-by-cell of extrusions) in the *PLASM* language; (b) The two-dimensional polyhedral complex obtained by the evaluation of the expression @2:(*Plan* && *Section*), represented as an exploded drawing

larly in describing the environment, the robot, and the motion, in off-line programming of robots.

Some examples of operators and expressions in the language follow. @:*Object* evaluates the boundary of *Object*; @*n*:*Object*, where *n* is a non-negative integer, evaluates to the *n*-skeleton of the complex which constitutes the representation of *Object*. For example, @0:*Object* evaluates to the set of vertices; @1:*Object* to the set of vertices and edges; and @2:*Object* to the set of vertices, edges and faces. The intersection of extrusions operator, denoted as &&, takes $p$ polyhedral complexes of dimension $(n-1)$ as arguments $(2 \leq p \leq n)$, and generates a polyhedral complex of dimension $n$ embedded in $\mathfrak{R}$. Thus if *Plan* and *Section* are two solid 2-complexes (with dimension 2 and embedded in $\mathfrak{R}$), then *Plan* && *Section* evaluates to the solid 3-complex whose cells are obtained by intersecting any pair of extruded cells of the arguments embedded into two coordinates planes of $\mathfrak{R}$. An exploded picture of the evaluated expression @2:(*Plan* && *Section*) is given in Figure 6.

Another important feature of the language is the evaluation of structures using an implicit progressive Boolean difference between the substructures. For example $S = STRUCT:<S_1, S_2, S_3>$ associates the name $S$ to the object whose cells are obtained by traversing the substructures $S_1$, $S_2-S_1$ and $S_3-S_2-S_1$. Notice that a structure is forced to accommodate only

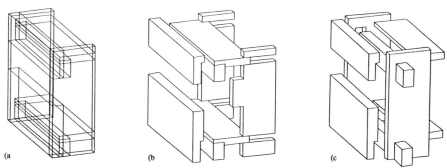

**Figure 7:** (a) A sequence (Beams, Roofs, Enclosures, Partitions) of isothetic (interpenetrating) pairs of parallelopipeds, for sake of simplicity assumed to be in the same coordinate space; (b) *PLASM* evaluated expression *STRUCT*: (Beams, Roofs, Enclosures, Partitions); (c) STRUCT: (Beams, Partitions, Enclosures, Roofs)

substructures with the same dimension. The geometric detail of such a traversal operation obviously depends on the sequencing of elements in the structure. Two examples of such operations are given in Figure 7.

## Acknowledgements

First of all I would like to thank Laura Moltedo, of the Board of the Italian Computer Graphics Association, for all her advice and for encouragement to write this paper. Many people have given a significant contribution to *Minerva*. I would also like to remember Carlo Cattani, Giulio Materlian, Mauro Mele, Maurizio Ramella, Maurizio Rosina, Alessandro Santarelli. Mauro Masia has developed the user interface, the structure subsystem, and continues to be responsible for system integration. The *Simple* $^n_X$ multidimensional modeler is being developed by Fausto Bernardini and Vincenzo Ferrucci. The preliminary work on the *PLASM* language has been done together with Claudio Sansoni. Most of the pictures in this paper have been prepared with *Minerva*. This work has been supported in part by the PF Edilizia of the Italian Research Council, under the contract CT CNR 90.01702PF64.

## References

[1] Bernardini, F.,. "Integration of Polynomials over $n$-Dimensional Polyhedra," *Computer Aided Design*, 23(1): 51–58, February 1991.

[2] Bernardini, F., V. Ferrucci and A. Paoluzzi, "Working with Multidimensional Polyhedra," Submitted paper. January 1991.

[3] Cattani, C. and A. Paoluzzi, "Boundary Integration over Linear Polyhedra." *Computer Aided Design*, 22(2): 130–135, March 1990.

[4] Ferrucci, V. and F. Bernardini, "*Simple$_X^n$*: User Manual and Implementation Notes. Parts 1 and 2," Technical Reports RR.90-06/17, Dipartimento di Informatica e Sistemistica, Univ. "La Sapienza," Rome, May 1990.

[5] Ferrucci, V. and F. Bernardini, "*Simple$_X^n$*: Un Modellatore Geometrico Multidimensionale," Proc. of *IcoGraphics '91*, 513–526, Mondadori Informatica, Milan, March 1991.

[6] Ferrucci, V. and A. Paoluzzi, "Sweeping and Boundary Evaluation of Multidimensional Polyhedra," *Computer Aided Design*, 23(1): 40–50, February 1991.

[7] Masia, M., G. Materlian, and M. Mele, "Approssimazione Poliedrica di Solidi Curvi." Proc. of *IcoGraphics '90*, 49–62, Milan, Feb 1990.

[8] Paoluzzi, A. and C. Cattani, "Simplicial-Based Representation and Algorithms for Multi-dimensional Polyhedra," manuscript, 1990, revised and extended version of the paper by C. Cattani and A. Paoluzzi, "Solid Modeling in Any Dimension," *Dip. di Informatica e Sistemistica*, Università "La Sapienza," Techn. Rep. 02–89, Rome, Jan 1989, revised May 1989.

[9] Paoluzzi, A. and M. Masia, "The Geometric Modeler Minerva," *Wheels for the Mind*, Apple University Consortium (Europe), 3(2):14–32, April 1989.

[10] Paoluzzi, A., M. Ramella, and A. Santarelli, "Boolean Algebra over Linear Polyhedra," *Computer Aided Design*, 21(8):474–484, 1989.

[11] Paoluzzi, A. and C. Sansoni, "Solid modeling of architectural design with PLASM language," To appear in the Proc. of *CAAD Futures '91*, Schmitt G. (Ed.), Zurich, July 1991.

[12] Paoluzzi, A. and E. Vietri, "Rappresentazione e Visualizzazione di Varietà e Campi su Varietà," Proc. of *IcoGraphics '91*, 541–552, Mondadori Informatica, Milan, March 1991.

# WORKING GROUP REPORTS

# Working Group Reports

The Computer Graphics and Education '91 conference was intended to be more than simply a collection of presentations on educational applications of computer graphics. Significant parts of the conference time were devoted to a set of of working groups, which were set up to go beyond the presentations and address some of the questions that the presentations raised. Each group was charged with examining some facet of computer graphics in education and with making appropriate recommendations dealing with that facet. The groups were formed based on a list of possible aspects, and each conference participant joined the working group that best represented his or her own interest in the field. The reports are shown in this section, and a list of participants follows at the end.

The working groups were set up to take advantage of the breadth of disciplines and experience found in the conference participants. Five groups were originally formed and two of these merged, yielding the following four groups:

- *Visual Learning.* This group explored the deficiencies in current learning environments and the opportunities to support visual learning in order to broaden students' learning opportunities.
- *Exploitation of Current Technologies to Improve Learning.* This group discussed various constraints on full utilization of current technology in instruction and suggested ways to make current computer graphics technology more effectively used in teaching.
- *Computer Graphics as a Tool in Teaching.* This group assessed the current state of computer graphics as an instructional tool and projected additional ways that computer graphics could support teaching in various disciplines.
- *Long Range Views of Computer Graphics and Education.* This group considered technological advances expected in the next five years that can be expected to have an impact on instruction.

# Visual Learning (Visual Literacy)

It is commonly understood that in order to acquire understanding, we need both logical and analogical thinking in about equal measure. The role of the teacher at all levels is to help students develop this understanding by providing them with the necessary intellectual building blocks, as well as the skills to be able to use them effectively. Some building blocks need to be symbolic and verbal; others need to be iconic and manipulative. There is, however, ample evidence that teaching often tends towards one of these polarities at the expense of the other. Hence some students become verbally literate and others become visually literate. Only rarely are these forms of literacy combined sufficiently well to allow students to rise to their fullest potential.

Although we stress that for a proper education, all forms of literacy need to be developed, the aim of this part of the report is to propose avenues that need to be explored in order to develop a broader *visual literacy*. We concentrate on the role to be played by the use of interactive computer graphics and multimedia as enabling technologies.

The value of computer graphics and computer animation for pictorial and diagrammatic illustration is already quite clear. Computer graphics has established its place in the day-to-day repertoire of presentation methods. By use of computer graphics painting or modeling systems, even the most manually unskilled of draftspersons can, with patience and some aesthetic judgement, produce images of a quality not previously achievable by non-specialists. This has presented domain experts with the possibility of including much more visual matter in their presentations, whether page-based or screen-based, than ever before. In all developed countries the use of computer graphics on television news, current affairs, and documentary programmes has allowed more direct, speedy, and cost-effective explanations of phenomena than were previously possible. Furthermore, this broadcasting use has given everyone, particularly the young, a familiarity with the appearance and value of this form of imagery. It is, therefore, not necessary here to repeat the advantages of computer graphics and computer animation as passive media for presentation of information either in printed or in screen form.

What this report considers, on the other hand, is the role for a more interactive and, indeed, more pro-active use of these media in teaching and learning. This role embraces items such as interactive electronic assistance for student-controlled learning and the simulated laboratory and workshop in all its forms. Such elements, which are discussed in detail below, have recently been made more feasible in day-to-day educational situations by continued improvements in the power/costs ratios of hardware and software. These improvements have given us economical means not only to store and retrieve at great speed vast quantities of information in the forms of words, sounds, and pictures, but also to compute and display in real time new information in at least some of these forms.

This is not to suggest that all technological problems have been solved. Especially in real-time simulation of both processes and objects, much more needs to be done before we can compute on the fly a fully-rendered animated scene, or simulate and illustrate in real time the development of time-based physical processes such as wind vortices around buildings. These tax even the most powerful of today's computers. But technological advances already on the horizon, and discussed in more detail elsewhere in the report, give us confidence that these things will be easily possible within the next five or so years.

We must, though, take at least two important factors into account in talking too enthusiastically about possible future advances, however desirable their outcomes might be. First, even in the developed world, not all countries and not all educational institutions in any country can invest in the newest technology to the same extent. There are great disparities in the facilities and resources available; important items of hardware and software that might be commonplace in one institution are very rare indeed in others. The unequal distribution is even more apparent when one considers the less-developed world.

Second, a great investment in computing, its related technologies, and above all, training has already been made in some educational areas and still has potential which cannot be lightly discarded. Our desire to throw off the old and put on the new is natural, and perhaps all of us would like to start again from scratch when we see what is currently available to the most advanced establishments. But this is at least as impractical in our institutional lives as it is in our personal ones. Thus not only must we look towards ways of exploiting the newest computing developments but also the old ones too.

In addition, because of the numbers of pupils and students involved, it is also important to take into account the vastly differing scale of investment requirements for educational computing technology at varying levels: preschool, primary, secondary, tertiary, and post-tertiary. However, were it possible to develop educational software which would be easily applicable, for example, to teaching aspects of physics, mathematics, or chemistry at a secondary school level in both Europe and the Americas, the burden of the initial hardware and software investment might possibly be offset by savings achieved from both the common understanding of the subject that would arise and the minimal, amortised cost of the educational packages involved.

Thus some investigation needs to be undertaken into where the most efficacious application of resources might be, both from the point of view of subject matter and of educational level. As part of this investigation we should look at whether it would be more useful for developments to take place as a series of small and possibly *ad hoc* efforts exploiting individual enthusiasms and abilities (that is, expanding the model that is being used at the moment), or as a world-wide integrated effort (as is the case of the Human Genome Project, an internationally coordinated attempt to catalogue the detailed structure of human DNA). Although the Genome Project, in common with similar international projects, has its critics, the advantage of this approach is that, possibly because of their propaganda value to governments involved, very often huge funding can result. The advantage of the *ad hoc* approach is that things often get done more rapidly and with less compromise than in the more formal approach.

Many of the problems and the possibilities opened up by interactive computer graphics and multimedia are hardly technological ones at all. They are educational and, possibly, attitudinal. Thus new pedagogical and organisational strategies will have to be developed if we are to move ahead. This will involve multidisciplinary research, experimental development, and training. Above all, it will mean much closer collaboration than has been the norm between teams of education theorists, computer graphics specialists, information designers, domain experts, and, if multilingual work is needed, linguists. It is very unlikely that the skills of just one person or even those of a small, single-discipline team would be sufficient to create the type of courseware that we envisage and which the new technology can support. Thus we should think of some of the excellent examples that have already been developed by such a route as illustrative of the possibilities and potential rather than as items to be judged in their own right. On the other hand, there is need both to collect and catalogue these examples and to investigate their more universal applicability from the pedagogical and the domain-knowledge point of view.

## Interactive Electronic Assistance for Student-Controlled Learning

This report assumes that interactive computer graphics has great promise for improving and focussing students' visual awareness. It should meet this promise in two ways:

- By providing congenial facilities for teachers and students to create, and examine interactively from many viewpoints, models of objects and physical processes;
- By presenting students with the tools for developing ideas in the form of sketches, diagrams, graphs, conceptual networks, and so on, so that they can improve their abilities to externalise. It is interesting that Leonardo, in common with many other Italian artists, did not consider sketches in his notebooks to be art-like pictures at all and referred to them as *penseri*, that is, "thoughts."

Evidence suggests that it is our differing ability to externalise concepts by means of drawing, sketching, and visual notes that distinguishes the visually literate from others, although the skills involved can be learned. This does not imply that to be visually literate, students have to be accomplished artists and draftspersons. On the contrary, externalisation can take many forms and even very skilled artists and designers often use a sort of personal pictorial but very simple shorthand, sometimes very difficult for others to understand, when they are developing or noting ideas. Of course, by their very natures, published examples from sketchbooks and notebooks are likely to have been chosen because they are at least slightly meaningful to others. Examination of the actual notebooks of artists, designers, and scientists makes even more cogently the point about the private and pictorially codified nature of most visual externalisation.

Current interactive computer graphics systems are extremely well-developed to meet the requirements of students and teachers for presenting finished graphical forms, but they are woefully inadequate when we want to use them for teaching visual note-taking and externalisation. Certainly the usual input tools do not even remotely match the congeniality and convenience of the pencil or pen. It is as if system designers had assumed that only "finished" graphics are useful and that we only use pictorial and diagrammatic means to convey information to others — rather than, as we have argued here, interactively to develop information for increasing our own understanding too.

It might, indeed, turn out that interactive computer graphics systems can never be conveniently used either for using or teaching the conventional sort of visual note-making and sketch-based externalisation of the type we are talking about here. The matter should be investigated not only to ascertain whether this is so but also to find out what alternative methods are appropriate to the new technologies and what new teaching strategies need to be developed in order to exploit them.

But perhaps the most important educational role for interactive computer graphics and multimedia systems with appropriate courseware is in assisting students with unsupervised learning. Certainly, the potential for helping students understand many concepts in mathematics and physics have already been well demonstrated. However, to enable these systems to work effectively without a teacher being present implies, among other things, the need for research into providing system-generated explanations when students run into difficulties. Human teachers provide explanations at three levels:

- Providing additional information to fill in apparent gaps in the students' knowledge,
- Correcting misunderstandings and misconceptions, and
- Suggesting hypotheses to allow students to develop frameworks for understanding new sets of facts and relating these to existing knowledge.

The first of these is easy to emulate in conventional computer-aided learning simply by releasing more and more stored information as required. The others are not at all trivial because, as we have learnt from research into the analogous case of explanation for expert systems, it is not sufficient to have a store of ready-made explanations to hand. Explanations have to be generated to suit the difficulties as they arise. This is a matter for artificial intelligence and natural language understanding and is further exacerbated if we require the explanations to be pictorial rather than textual.

Thus another avenue to be explored is the way in which cogent explanations of the three types listed above can be provided by the systems when the students need them. For this to happen, it is probable that the systems will have to embody at least five models:

- A model of the domain under consideration,
- A model of the students' learning process (or processes),
- A "bug" model, a model of the sort of mistakes that are likely to occur,

- A model of graphical and textual explanation processes, and
- A model of assessment procedures (to help evaluate students' progress).

Few, if any, of these models currently exist in computable form. Most of them will be subject dependent and will, of course, vary according to the level of subject difficulty. More than anything, though, they will need a great deal of research to ascertain their details and applicability.

In summary, then, the role of interactive computer graphics and multimedia in increasing students' visual literacy, and hence understanding, appears potentially limitless. Realising that potential, however, will not be easy. It will require research: pure, applied, and experimental development, at all levels. Above all, it will require cooperation across many disciplines: not only the disciplines actually being taught but also those of education theory, cognitive science, computing and computer graphics, and visual information design. Whether this work is best done at an international, national, or local scale needs to be examined, but it certainly needs to be done because the educational experience cannot be considered complete unless visual literacy plays its fullest part.

# Exploitation of Current Technology to Improve Learning

The goal of this working group was to examine how to take immediate advantage of recent technology developments so as to improve the learning process. Of particular relevance, of course, are advances in graphics technology: today the educational workstation is typically a PC or UNIX workstation, with graphics as an integral part.

There is a long history of attempts to provide computer-support tools for education (see for example the paper in this volume by Levrat). Some of these attempts have been major projects such as PLATO; others have been small research projects. Somehow the promised changes in the styles of learning have not been fulfilled; traditional styles continue to predominate.

This working group was established to examine the reasons for this poor take-up of computer-aided education and to seek the best means of promoting its use. This effort was given some momentum by the presentations at the conference which showed the new opportunities offered by multi-media approaches, involving graphics, text, and sound.

## Constraints on Use of Current Technology in Education

A number of reasons were given for the resistance to the introduction of computer-support tools. These can be broadly categorised as issues concerning people, technology, and economics.

PEOPLE CONSTRAINTS:

1. There is an innate conservatism amongst many people in the educational community. This barrier to change can be difficult to overcome and certainly inhibits experimentation in new teaching methods.

2. In many instances there is a lack of reward to the individual for effort put into the development and use of educational software. The university system in many countries places greater emphasis on research than on teaching. There is limited recognition of teaching performance, and hence little motivation. (Note this is not universally true; for example, Encarnação pointed out that teaching skills are properly acknowledged in Germany).

3. A major effort, in time, intellect, and resources is required to create a first-class educational software tool. Van Dam, who should know, remarked

at the conference that the effort was greater than that involved in writing a textbook on the same subject. Studies at Cornell University in the U.S. suggest 100 to 400 hours of effort are required to create a one-hour, on-line instruction module.

4. Most lecturers like to impart a personal touch to the material they are presenting. Unfortunately many pieces of educational software are inflexible and prevent a lecturer customising the software.

*TECHNOLOGY CONSTRAINTS*
1. The rapid developments in computer hardware have led to software that is soon out-of-date. It has proved difficult to create software that is scalable in the sense of being able to take advantage of new technology. This works against long-term use of the software and is a disincentive to its creation in the first place, bearing in mind the remarks earlier on the effort involved.

2. The wide range of computing platforms used in education has led either to software dedicated to one platform or software which uses the lowest common denominator of many platforms — neither being satisfactory. For example, it was noted by Owen that HyperCard-like systems, available on PCs as well as the Apple Macintosh, are not available on UNIX workstations at the present time.

3. There is a lack of standards in some areas; in others, the slow gestation period of standards has discouraged their use.

4. There is sometimes conflict between technology and pedagogy. The technology should only support, not dictate. Some tools have overplayed the technology and put little thought into the teaching and learning aspects.

*ECONOMIC CONSTRAINTS*
1. The private sector does not appear to regard educational software as a commercial proposition. This includes both publishers and computer manufacturers.

2. Funding for educational software has been typically short-term, resulting in good prototypes but little production quality material. There has been no support and maintenance and no on-going development. The products have, therefore, tended to wither.

3. Universities and higher education generally have been slow to fund the support technology needed. Lecture rooms tend to be poorly equipped in terms of projection systems — which admittedly are not cheap, with $10,000 cost for a colour video projector.
4. Software licencing policy often works against providing tools on each educational workstation. The need for educational site licences is imperative.

## *The Way Forward*

The model of supercomputer centres in the U.S. has been very successful. Major funding to establish these centres on a long-term basis has proved a catalyst to the use of supercomputers in the U.S.; they are now widely used for research throughout the university system.

A similar model has been proposed for educational software. A recent workshop was held at NCSA at the University of Illinois at Urbana-Champaign. This made four recommendations:

1. To increase the number of teachers who use computer-support tools (by teacher-training initiatives),
2. To support the development of tools and materials (by initiating development and support activities),
3. To revise curricula to integrate computational science methodology, and
4. To provide access to advanced tools through networking.

The working group broadly concurred with these recommendations. In particular, the need for ongoing support was recognised, to avoid the mistakes of the past. It was felt this was best done by the establishment of a small number of properly funded centres. The group therefore recommended that centres for educational software should be established. These would receive long-term, public funding to develop, support, and maintain educational software. The tools would be created by interdisciplinary teams: these teams would comprise

- a specialist in the subject concerned, to ensure sound pedagogical content,
- a software engineer, to ensure high software quality and ease of porting to different platforms, and
- a graphic designer, to advise on presentation.

This model has been successful in the area of scientific visualisation and should carry over well to education. It is expensive in manpower but becomes economic through widespread use of the tools developed.

International cooperation is seen as an important aspect to avoid duplication of effort and to address the issue of language and cultural differences. It was noted that there have been European Community projects in the area of education (see the paper by Hornung in this volume) and new initiatives are expected to be announced soon. In the U.K., the "Computers in Teaching" initiative has led to the creation of a large number of small centres, one per subject area.

## Current Products

The working group also noted a number of tools currently available for use in developing teaching software.

1. *HyperCard.* The HyperCard system, originally introduced on the Macintosh but also available now on PCs, allows teaching material to be developed as a set of "cards" containing graphics and text, with links established between words and pictures on a card and other related cards. This system is being successfully used by many educators, though concerns exist over the card metaphor and the close coupling of the links to the material itself. This is described in the paper by van Dam.

2. *Guide.* The Guide system, available on both PC and Macintosh, provides the linking capability of HyperCard without the constraint of the card metaphor. It is essentially a file-and-window-based system: information is stored as a set of files containing graphics and text; the files may be browsed in a window environment, with several windows being shown on-screen at any time. This will be used in the SIGGRAPH Hypergraph project, described in the paper by Owen.

3. *NeXT Computer.* Conference participants were impressed by the demonstration of the NeXT computer by Richard Phillips during his presentation. He showed the use of MediaView, a "multi-media, digital publishing system" supporting text, graphics, audio, and video. A recent project had implemented several chapters of the Foley et al textbook on computer graphics: the text literally came alive as Dr. Phillips showed how the algorithms could be animated, voice-overs from the authors added, and personal notepads for private learning incorporated. This demonstration gave a glimpse of what would be available to educators in the very near future as a tool.

4. **Computer Aided Lecturing.** The presentation by Jacques Raymond showed a very useful tool for use in lecturing. His system, "Le Prof," provided an "electronic blackboard" with computer simulation of blackboard, overhead projector, and 35mm slides. It runs on a PC and can be used in conjunction with a video projector or an LCD unit placed on an overhead projector.

## Conclusions

It is to be hoped that the potential of computer support tools for education can finally be realised. It may well be that the multi-media technology advances will provide the impetus to overcome the constraints that have proved an obstacle in the past.

The role of the lecture will surely change. There seemed to be a common view from delegates of different nationalities that students are poorly motivated to learn and that the present lecturing system was partly to blame. The novel system of Raymond would provide a more attractive lecturing environment, but long term there should be a move to more "self" learning by the student from computer-based materials. The human touch is still needed to give inspiration and motivation to the students, but the way this is given will change.

An important aspect, recognised by the working group, is the need to review the way students are assessed under the changing styles of learning.

As the trend continues to individual learning rather than mass teaching, so the need for individual student workstations becomes important. The recent announcement of portable UNIX workstations is an important development.

The main conclusion of the working group is that a radical change in attitude is needed within the higher education system to give greater importance to the development of educational courseware. The establishment of support centres is a necessary step if this development is to be properly coordinated and the impetus sustained.

# Computer Graphics as a Tool in Teaching

This working group included seventeen people representing four continents and fourteen countries. Our first task was to assess the current state of the use of computer graphics as a teaching tool. To do this each person (or persons when there was more than one person per country) gave an informal estimate of the use of computer graphics in education in the areas of computer science, science, engineering, art and design, and others. The scale was from 1 to 10, with 1 representing no use at all and 10 representing optimal useage. To preserve anonymity, countries are given only as numbers and are not in alphabetical order (Table 1). As can be seen from the table, no country is now even close to using computer graphics in an optimal manner, and most countries use it in a minimal fashion. One problem is the disparity in acessability and cost of machines. For an IBM AT class machine, the cost ranged from about two weeks average wages to almost ten years' wages, with most of the countries grouped at the lower end.

The group agreed that at present there is little transfer of educational software, so that most of the software is developed in the same country in which it is used We also looked at the current platforms in use and the distribution was as follows: IBM compatibles 60 percent, Macintosh 30–40 percent, and UNIX systems 0–10 percent.

| Country | Comp. Science | Science | Engineering | Art & Design | Other |
|---|---|---|---|---|---|
| 1 | 4 | 6 | 7 | 4 | 1 |
| 2 | 3 | 4 | 5 | 1 | 1 |
| 3 | 3 | 3 | 5 | 1 | 2 |
| 4 | 3 | 2 | 3 | 1 | 1 |
| 5 | 2 | 2 | 3 | 4 | 1 |
| 6 | 4 | 2 | 4 | 2 | 1 |
| 7 | 4 | 6 | 6 | 5 | 1 |
| 8 | 2 | 3 | 3 | 4 | 1 |
| 9 | 2 | 5 | 7 | 7 | 1 |
| 10 | 3 | 3 | 3 | 6 | 3 |
| 11 | 2 | 2 | 2 | 2 | 2 |
| 12 | 4 | 5 | 6 | 5 | 1 |
| 13 | 5 | 3 | 7 | 4 | 2 |
| 14 | 2 | 3 | 2 | 1 | 1 |

**Table 1**

## Transfer to Education

The second major topic was the educational applications of computer graphics. The group agreed that computer graphics ideally should be used for every topic and at every level. However, we decided to restrict our discussion to the university level and look at those needs and uses. We agreed that the use of computer graphics leads to a deeper and better (more intuitive) understanding of the subject matter. It should be used both for demonstration purposes, such as lectures, and for out-of-class student use. We thought that students would not only gain a better understanding of the material, but the use of computer graphics would allow the courses to cover more material in a given time. This is important, since there is always more material to cover than possible in the allotted time, and the amount of material is constantly increasing.

The group thought that the use of computer graphics would impact on teaching and learning styles in the following ways: for professors it will require more initial effort for class and course preparation and would require new methods for preparing lectures; it will require the development of supporting tools for lectures; it may allow for fewer lectures; courses will be more student-centered and will allow for more adaptation for individual student needs. The students will use the software outside the classroom to deepen their understanding and improve their intuition about the subject matter. The courses will become more self-paced.

## Project to the Future

The use of computer graphics in teaching will lead to new ways of thinking or doing. It will allow for more powerful and symbolic manipulations, and thus it will allow for a higher level of representation. For example, we will be able to manipulate objects (data plus operations) rather than just data. It will allow us to visualize ideas and concepts otherwise difficult to see in reality ("to see the unseen"). Students will be able to simulate real-life experiences that might be too dangerous or expensive. An example of this is simulation of chemistry experiments where the chemicals are very expensive or the chemical reaction is dangerous. Computer graphics will allow us to augment or prototype the making of physical models. It will allow us to perform medical simulations ("you can dissect a frog without the frog minding"). In some medical schools there is a shortage of cadavers, and so surgeons might simulate operations using computer graphics. Note that we are not suggesting that the simulations completely replace the actual procedure, but the student can practice first on the computer before trying it for real.

Computer graphics can also be used to stimulate and motivate students. All the faculty in the group agreed that their students tend to be unmotivated and somewhat lazy. A good example of the power of computer graphics to make otherwise dry subjects interesting is the *Mathematics!* videos produced by Tom Apostol and Jim Blinn. Computer graphics will lead to new insight and a discovery of new aspects of the subject material.

Computer graphics will lead to bridges between disciplines. It will become necessary for people to learn more about visual thinking and about the effective presentation and display of visual information. This will lead to the introduction of aspects of art and design and cognitive psychology into other disciplines. Thus, computer graphics can act as a bridge between the arts and the sciences. An example of this is the use of computer graphics to represent dance-notation languages used to record and develop choreography.

*Implementation Issues*

To implement the full use of computer graphics in education there must be cooperation between professional societies and software developers to determine just what types of tools are needed. The needs of the educators must be assessed and then the appropriate tools must be developed and distributed.

We must take advantage of economies of scale; the developed tools must be widely distributed so that individual copies are relatively inexpensive. For example, in the PC market, developers may expect tens of thousands of sales and so most software products cost only around $100; in the UNIX workstation market, the equivalent product might sell only a few thousand copies and so the individual copy price might be around $1,000. To achieve wide distribution and economy of scale, the tools must be easy to use and easy to integrate into the classroom.

Along with finished applications, it is also necessary to provide tools so that educators can easily create their own applications. Even the application programs should be easy to modify and adapt for different educational environments. The developed tools must use standards for interoperability and portability between systems. Too much useful CAI software has disappeared when the computer system on which it was implemented has become obsolete. The tools should be adaptable to take advantage of the increasing power of the computer systems as they evolve from the present 3M machines to 3G and even 3T machines.

The group felt that higher level tools should be developed so that the educators themsleves could tailor them to specific applications or methods of practice. One stated goal was to reduce the development time for excellent CAI software by a factor of 10 to 100. Some specific tools mentioned were general testing tools, tools for interactive discovery, tools which provided the student with a problem solving environment, tools for high-quality presentation graphics, and for computer graphics instruction, $n$-dimensional modeling (at least four-dimensional for easy animation). The tools should incorporate aspects of artificial intelligence, as in intelligent tutoring systems. They should generate problems and then provide help in solving the problems. If the tools are going to be used internationally then they should have multi-lingual support, that is, versions for different human languages.

To accomplish the goal of the production of these tools requires a major international effort. The first, and easiest, task is to convince teachers of the need. This is already partly done. The second task is to convince university administrators, national funding agencies, and international funding agencies of the need. Commercial developers should also participate in the development effort.

# Long Range Views of Computer Graphics and Education

The group sketched a hypothetical educational workstation for 1996. The hardware platform, the low end for a major graphics manufacturer, is expected to have a 200-MIPS processor with 100-MFLOPS capability, 64-Mbyte RAM, and 1-Gbyte disk, be able to render 250,000 triangles per second (texture-mapped, alpha-buffer, stereo), and sell for $5,000 to $10,000 (in 1996 dollars). Its communications will consist of a FDDI (100 Mbits) local-area network connected to a similar wide-area network with faster trunks (one Gbit), with operational speeds significantly slower than the local-area network due to traffic. Its input capacities will include six degrees-of-freedom devices: data glove, head tracker, and other virtual-reality devices; gesture recognition; voice recognition; force feedback; and tactile feedback; all at commodity prices (but these are not included in the price above). It will also include CD-quality stereo sound output.

Its basic software platform will include a multi-processing, multi-tasking, real-time operating system. TCP-IP (the FORTRAN of communications) will be the software basis for communications, having won out over other schemes. It will include an object-oriented development environment and a client interface to an object-oriented database. This database contains the application model and any geometric models and is designed to support real-time interaction and incremental computation, but renderers are allowed to cache whatever and however they want.

## *Graphics Software Prospects*

Current graphics software is sorely inadequate. The graphics domain must shift from subroutine libraries to an object-oriented framework, relying on persistent databases, graphic object libraries, and a visual programming environment. There will be a variety of primitives and operations traditionally associated with modeling and physics but only a small subset of which is now well understood and generally used. Motion, collision detection, deformation, and the like will be standard methods optionally defined for objects. We will content ourselves with rough approximations, just as Phong shading is a rough approximation to the physics. It's too early for a "unified field theory" of this graphics environment, but a rich set of ideas would include:

- A quality dial which allows the user to choose the order of approximation.
- A choice of renderers, related to the quality dial.
- Much more complex specification of object attributes will be available to the user, either from a graphical user interface directly (setting color and the like) or via callback from the simulation engine (for example, hook the color to an internal variable of the simulation).
- The line between the graphics framework and specialized computation will change as the latter is better understood and accepted into general practice. For example, in the far future we expect that the object-oriented database will be enriched with content-oriented retrieval on the various media.
- The emergence of multimedia is beginning to affect how we think about graphics and will doubtless affect the hardware and software environment in ways not yet apparent.

The question was raised whether or not rendering interface specifications are important. How sharp is the line separating rendering from the rest of the graphics software?

*Application Areas in the Framework*

There are a number of application areas that may be moved over the line into the graphics environment in the next five years. We assume each area will be a closed domain; for example, constructive solid geometry on volume-enclosing primitives yields volume-enclosing primitives. Subsequently, we may be able to extend operators to work across domains, so that, for example, deformation may be defined for both images and sounds. Among these application areas (with operations in parentheses) are:

- Solids modeling (booleans).
- Image processing (subset useful for texture mapping, such as filtering, scaling, warping, and the like).
- Volumetric modeling (booleans, rendering, and "volume" processing in analogy with two-dimensional image processing.)
- Pseudo-physics: mechanics, dynamics, constraints, finite element analysis, and more (transformations, forces, torques, collision detection on solid objects with mass).
- Time modeling: three-dimensional graphics plus time. All geometric and

appearance attributes and all operations are intrinsically time-varying. This time variation can be specified by direct manipulation, graphs, procedures or callbacks, and other techniques.

There is clearly a need for national initiatives to develop the software described in these sections.

## *Input Handling*

The twenty-year-old input models currently in GKS and PHIGS are wholly inadequate. At the lowest level there will be an extensible event record with time-stamping, but the system will permit the rapid building of higher-level interaction tools, essentially multi-dimensional widgets with arbitrarily complex behavior. In particular, there will be a trend to encapsulate application-specific semantics in widgets to provide immediate, application-dependent feedback during an interaction, such as automatic spell checking while you type. This will blur the line between input and computation, just as the distinction between computation and display is blurred as described earlier. The principle involved here is incremental versus batch design-rule and error checking. There is some controversy here: do current paradigms suffice?

## *Transfer to Education*

By creating the best environment for research, we also provide the most effective one for education. For example, the creation of electronic books and virtual laboratories will be made easier and much more powerful. This multi-media presentation system will be part of a larger educational environment including things such as intelligent tutoring systems, lying outside the framework described above.

## *Additional Notes and Comments*

The range of discussion in this working group is probably well represented by a question it raised: Is graphics a good word, or should we invent a new word for this multimedia stuff?

There were a number of comments about video developments. JPEG (Joint Photographics Expert Group) and MPEG (Motion Picture Expert Group, who built on the JPEG work) image compression systems are tuneable to the degree of loss. ISO and ANSI together have handled the JPEG standard and will also handle MPEG. Fluent Technologies now produces an ISA-bus compatible, multi-media board set including scaleable (see below) full-motion video, JPEG compression, and multiple video windows. By 1996, we can foresee MPEG that includes temporal coherence, known as frame coherence in graphics. In addition, there is the question of scaleable video: an adaptive video system. There is a server which handles video signals to the client. There is a graceful degradation to the quality of the image, the size of the image, or the depth of the image. This notion is currently not incorporated in JPEG or MPEG. It also creates a digital video standard which transcends any specific format (NTSC, PAL). Scaleable video then is orthogonal to the issue of digital or analog HDTV: one is a digital format, the other is a display format. The solution should be present well before five years from now.

The group raised the issue of synchronization of video and audio because there is now no standard for incorporating this into a multi-media database. This is a difficult problem. HyTime is an ISO standards group which started out looking at ways to incorporate musical information in a document for use with SGML documents. It's been expanded to a wider spectrum of documents, such as video. Other issues include multi-media mail standards and desk-top video conferencing. This latter is important because computer-supported collaborative work (CSCW) will involve real-time video and audio in shared windows, using this technology.

Footnote:

The April and July 1991, issues of the *Communications of the ACM* are special issues on interactive digital multi-media. They include articles on scaleable video, HyTime work, JPEG, MPEG, and the like. The April issue focusses on standards and the July issue on applications.

# Working Group Participants

*Visual Learning*

Hermann Härtel (FRG)
Joseph Hardin (US)
Michael Kitson (Australia)
John Lansdown (UK), chair
Barbara Mones-Hattal (US)
Ken O'Connell (US)
Jonathan Taylor (UK)
Thomas West (US)

*Exploitation of Current Technology*

Ken Brodlie (UK)
Judith R. Brown (US)
Pere Brunet (SP)
Morris Firebaugh (US)
Mario Rui dos Santos Gomes (PT)
Susan Laflin (UK)
Bernard Levrat (CH), Chair
Mike McGrath (US), Chair
Jacques Raymond (CA)
Nora Sabelli (US)
Wolfgang Strasser (FRG)
James Ver Hague (US)

*Computer Graphics as a Tool in Teaching*

Vladimir Anischenko (USSR)
Dolors Ayala (SP)
Daniel Crespo (SP)
Ana Cruces (MX)
Edo Dooijes (NL)
Niko Guid (YU)
Ahmad Nasri (UAE)
Adele Newton (CA)
Shogo Nishida (JP)
G. Scott Owen (US), chair
Alberto Paoluzzi (IT)
Norman Soong (US)
José Teixeira (PT)

*Long-Range Views*

Charlie Gunn (US), Chair
Christoph Hornung, (FRG)
Richard L. Phillips (US)
Andries van Dam (US)
José Encarnação (FRG)

# Participant List

Vladimir Anishchenko
School of Information Systems
University of East Anglia
Norwich, NR4 7TJ, U.K.

Dolors Ayala
Dept. Llenguarges i Sistemes Informatics
E.T.S.E.I.B.
Avda. Diagonal, 647 8º Pl
08028 Barcelona, Spain
*ayala@lsi.upc.es*

Judith R. Brown
Weeg Computing Center
The University of Iowa
Iowa City, IA 52242 USA
*jbrown@umaxa.weeg.uiowa.edu*

Ken Brodlie
School of Computer Studies
University of Leeds
Leeds LS2 9JT, U.K.
*kwb@des.leeds.ac.uk*

Pere Brunet
Dept. Llenguarges i Sistemes Informatics
E.T.S.E.I.B.
Avda. Diagonal, 647 8º Pl
08028 Barcelona, Spain
*brunet@lsi.upc.es*

Donna Cox
NCSA
University of Illinois
405 N Mathews Ave
Urbana, IL 61801 USA
*dcox@ncsa.uiuc.edu*

Daniel Crespo
Dept. de Fisica Aplicada
Universidad Politecnica de Catalunya
Jordi Girona, 31 edificio CH
08028 Barcelona, Spain

Ana Lilia C. Laureano Cruces
Univ. Autónoma Metropolitana-
 Azcapotzalco
Dpto de Sistemas
San Pablo 180 Col. Reynosa Tamps.
02200 México D.F., México

Steve Cunningham
Computer Science Department
California State University Stanislaus
Turlock, CA 95380 USA
*rsc@altair.csustan.edu*

Edo H. Dooijes
Mathematics & Computer Science
University of Amsterdam, Kruislaan 403
1098 SJ Amsterdam, The Netherlands
*edoh@fwi.uva.nl*

José Encarnação
Graphische Datenverarbeitung
Wilhilminenstrasse 7
D-6100 Darmstadt, Germany
*jle@agd.fhg.de*

Morris W. Firebaugh
Applied Computer Science Dept.
University of Wisconsin - Parkside
Kenosha, WI 53141-2000 USA
*morris@vacs.uwp.wisc.edu*

Mario Rui Fonseca dos Santos Gomes
Rua Alves Redol, 9, 2º D
1000 Lisboa, Portugal
*mrg@inesc.inesc.pt*

Niko Guid
Department of Computer Science
University of Maribor
62000 Maribor,
Smetanova, 17, Yugoslavia
*guid@uni_mb.ac.mail.yu*

Charlie Gunn
Geometry Supercomputer Project
1200 Washington Ave. S.
Minneapolis, MN 55415 USA
*gunn@geom.umn.edu*

Hermann Haertel
Inst. für die Paedagogik
der Naturwissenschaft
Universität Kiel
Olshausenstr 62
D-2300 Kiel, Germany
*haertel@nw-didaktik.uni-kiel.dbp.de*

Joseph Hardin
NCSA
405 N. Mathews Ave.
Urbana, IL 61801 USA
hardin@ncsa.uiuc.edu

Christoph Hornung
FhG-AGD
Wilhilminenstrasse 7
D-6100 Darmstadt, Germany
hornung@agd.fhg.de

Roger Hubbold
Department of Computer Science
University of Manchester
Manchester M13 9PL, U.K.
hubbold@cs.man.ac.uk

Christopher Jones
CERN, European Labs for Particle Physics
CN Division
1211 Geneva 23, Switzerland
chris@cernvm.cern.ch

Michael Kitson
43 Ardrie Road
East Malvern, Victoria, Australia 3145
art123h@monug.cc.monash.edu.au

Susan Laflin
School of Computer Science
University of Birmingham
Edgbaston, Birmingham B15 2IT, U.K.
laflins@bham.ac.uk

John Lansdown
50-51 Russell Square
London WCIB 4JP, U.K.
john17@clus.mx.ac.uk

Bernard Levrat
Université de Genève
Services Informatique
24, rue du Genéral Dufour
CH-1211 Genève 4
levrat@uni2a.unige.ch

Mike McGrath
Engineering Department
Colorado School of Mines
Golden, CO 80401 USA
mmcgrath@mines.colorado.edu

Barbara Mones-Hattal
Art and Art History Department
George Mason University
4400 University Avenue
Fairfax, VA 22030 USA
bhattal@gmuvax.gmu.edu

Ahmed Nasri
UAE University
Faculty of Science
Dept. of Math & Computer Science
P.O. Box 17551 AI-AIN
United Arab Emirates
fax: +971-3-671291

Isabel Navazo
Dept. Llenguarges i
Sistemes Informatics
E.T.S.E.I.B.
Avda. Diagonal, 647 8º Pl
08028 Barcelona, Spain
navazo@lsi.upc.es

Adele Newton
Manager, University Program
Alias Research
110 Richmond Street East
Toronto, Ontario M5C 1P1, Canada
anewton%alias@csri.toronto.edu

Shogo Nishida
Department of System Science
Central Research Laboratory
Mitsubishi Electric Corporation
Tsukaguchi-Honmachi 8-1-1
Amagaski, Hyogo 661, Japan
nishida@sys.crl.melco.co.jp

Ken O'Connell
Fine and Applied Arts
The University of Oregon
Eugene, OR 97403-1206 USA
oconnell@oregon.uoregon.edu

G. Scott Owen
Mathematics & Computer Science
Georgia State University
Atlanta, GA 30303 USA
matgso@gsusgi1.gsu.edu

Alberto Paoluzzi
Dip. di Informatica e Sistemistica
Università "La Sapienza"
Via Buouarroti, 12
00175 Rome, Italy

Jaume Pagés
Universidad Politecnica de Catalunya
Gregorio Marañon s/n
08028 Barcelona, Spain

Dick Phillips
Los Alamos National Laboratories
P.O. Box 1663 MS B272
Los Alamos, NM  87545 USA
rlp@lanl.gov

Jacques Raymond
Dept. Informatique
Université D'Ottawa
34 George Glinski
Ottawa, Ontario  K1N 6N5  Canada
jarsl@acadvm1.uottawa.ca

Gerhard Rossbach
Springer-Verlag Heidelberg
Hergerstrasse 17D
6900 Heidelberg, Germany
ROSSBACH@SPINT.CompuServe.COM

Nora H. Sabelli
NCSA
605 E. Springfield
Champaign, IL  61820 USA
nsabelli@ncsa.uiuc.edu

Norman L. Soong
Mathematical &
Computer Sciences Dept.
Mendel Hall, Room 171
Villanova University
Villanova, PA  19085 USA
soong@vuvaxcom.edu

Wolfgang Strasser
Wilhelm-Schickard-Institut
für Informatik
Auf der Moyenstelle 10
C9 Universität Tübingen
7400 Tübingen, Germany

Jonathan P. Taylor
Dept. of Computer Science
Queen Mary and Westfield College
Mile End Road
London E1, England
jtaylor@cs.man.ac.uk

José Carlos Teixeira
Grupo de Métodos e Sistemas
Gráficos Dept. Matemática
Universidade de Coimbra,
Ap. 3008
3000 Coimbra, Portugal
teixeira@ciuc2.uc.rccn.pt

Joan Truckenbrod
Art and Technology Area
School of the
Art Institute of Chicago
37 South Wabash
Chicago, IL  60603  USA

Andries van Dam
Computer Science Department
Brown University
Providence, RI  02912 USA
avd@cs.brown.edu

James C. Ver Hague
College of Fine and Applied Arts
Rochester Institute of Technology
One Lomb Memorial Drive
Rochester, NY  14623 USA

Thomas G. West
6622 32nd Street NW
Washington, DC  20015 USA
4139461@mcimail.com

# Appendix II

## International Programme Committee

This list includes several individuals who also participated in the conference programme. Their addresses have been omitted here.

Judith R. Brown

Ken Brodlie

Pere Brunet

David Clark
Univ. of London Audio Visual Centre
North Wing Studios
Senate House
Mallet Street
London WC1E 7JZ, England

Donna Cox

Steve Cunningham

Jose Encarnacao

Bianca Falcidieno
Instituo per la Matematica Applicate CNR
Via Alberti 4
16132 Genova, Italy

Robin Forrest
School of Information Systems
University of East Anglia
Norwich NR4 7JT, U.K.

John Gero
Dept. of Architectural Sciences
University of Sydney
Sydney, N.S.W. 2006, Australia

Marion Günther
Graphische Datenverarbeitung
Wilhilminenstrasse 7
D-6100 Darmstadt, Germany

Wendy Hall
Department of Computer Science
University of Southampton
Southampton SO9 5NH, England

Bob Hopgood
Informatics Department
Rutherford Appleton Lab.
Chilton, Didcot
OXON OX11 0QX, U.K.

Roger Hubbold

Bernd Kehrer
Wilhelm-Pick-Universität Rostock
Sektion Informatik
Albert-Einstein-Str 21
2500 Rostock, Germany

Lars Kjelldahl
Royal Institute of Technology
Nada, Kth
S-10044 Stockholm, Sweden

Tosiyasu Kunii
Dept. of Information Science
Faculty of Science
The University of Tokyo
7-3-1 Hongo, Bunkyo-ku
Tokyo 113, Japan

Leon Lastra
IIE, Unidad de Computo
Apdo Postal 465, CP 62000
Cuernavaca, Mor., México

Michel Lucas
Université de Nantes
2 Chemin de la Houssinière
44072 Nantes Cedex, France

## International Programme Committee

Mike McGrath

Laura Moltedo
JAC CNR
Via del Policlinico 137
00161 Rome, Italy

Barbara Mones-Hattal

G. Scott Owen

Frits Post
Delft University of Technical
   Mathematics and Informatics
Julianlaan 132
2620 BL Delft, Netherlands

Madalena Quirino
Universidade Nova de Lisboa
Departamento de Informatica
Quinta da Torre
2825 Monte da Caparica, Portugal

Juris Reinfelds
Department of Computer Science
New Mexico State University
Las Cruces, NM 88003-0001, USA

Wolfgang Strasser

José Carlos Teixeira
Grupo de Métodos e Sistemas
Gráficos Dept. Matemática
Universidade de Coimbra, Ap. 3008
3000 Coimbra, Portugal

Tetsuo Tomiyama
University of Tokyo
Dept. of Precision Machine Engineering
Faculty of Engineering
Hongo, 7-3-1, Bunkyo-ku
Tokyo 113, Japan

Andries van Dam

Jacques Weber
Laboratory of Computational Chemistry
University of Geneva
30 Quai E. Ansermet
1211 Geneva 4, Switzerland

Robert S. Wolff
Apple Computer, Inc.
609 Durwood Drive
La Canada, CA 91011, USA

# Index

Abstracting, 130
Abstract Interface, 146-147, 155-156
Abstract windows, 155
Academia, *see Schools,* Universities
Access, telecommunications, 187
ACM SIGGRAPH Education Committee, 76
ActionMedia, 26
Adaptability, courseware, 203, 207
Advanced Learning Technology (ALT), 62
Aesthetics, 130
African art, 106
Agents, 18-19
Algorithms
  efficiency of, 152-153
  graphics, 68-69, 217-225
  MediaView and, 30-31
  real-time animation, 20
  simulation of, 217-225
  study of, 68-69
Alias Research, 211, 215
Alias software, 111
Allman, John, 132
Ameritech, 175
Analogical thinking, 80, 130, 244
Analog video, 27
Anatomical illustrations, 189
Anchors, hypermedia, 14
Animation, 190
  algorithms for, 20
  HyperGraph and, 75-76
  light and, 75-76
  realtime, 12, 19-24
  REL and, 192-197
  3D, 20-21, 110-113
  user-controlled, 19-24
Annotation links, 71, 73
ANSI, 261-262
Apple, 15, 26
Application environments, 55-56
Application interface design, 123-124
Application links, 71, 75

Apprenticeship, 93
Architecture, 215, 227-228
Arithmetic, 92
Art, 106
  CAID and, 211
  collaboration in, 176-177
  computer graphics and, 112
  individualism in, 190
  interactive, 185-186
  mathematics and, 194
  science and, 257
  teleconferencing and, 174-177, 179-187
Art Center College of Design, 215
Art History, 106
Artificial Intelligence, 18-19, 70, 248-249
Artists
  scientists and, 189
  visual thinking by, 129-131
Assessment, 55
Association, 37-38
Associative methods, 44
Associative trails, 14
ATM, 51
Attention, 40
AT&T, 175, 184
Audio, *see Sound*
Authoring
  animation, 20
  automating, 19
  hypermedia, 12, 67-68
  languages for, 208
  multimedia, 12
  systems for, 13, 67-68
Automatic authoring, 19
Automatic programming, 209
A\UX, 15

Backus Normal Form (BNF), 15-16
BALSA, 20
The Banishment of Paperwork, 10
Baumgart's winged-edge, 231
Beckerman, Christoph, 107

Beckman Institute, 192
Biological illustrations, 189
Blackboard metaphor, 103-113, 161-162, 165, 170, 254
BNF (Backus Normal Form), 15-16
Books
 branching, 44
 electronic, 9-24
 vs. experience, 93, 99
 Hyper, 202, 209
 lifetime of, 209
 paper, 9-10, 44
 *see also* Textbook
Boolean operations, 234-238
Boundary representation, 231, 235-236
Bourbaki group, 101
Brain growth, 94
Branching books, 44
Break points, 154
Broadband networks, 51
Brown, Paul, 211
Building design, 227-228
Bulletin boards, 10
Bush, Vannevar, 13-14, 17, 37, 43, 104

C, 143
CAD, 70, 135, 185, 227
CAI, *see* Computer Aided Instruction
CAID, 211-215
CALS initiative, 24
Camera model, 234
Cards, hypertext, 67
Caricature, 195
Cartographic data, 109
Cataloguing, image, 41
CD-I (Compact Disc Interactive), 25
CD-ROMs, 11, 27, 28, 41, 66
 Greek text on, 204
Cell complex, 236
Center for Integrative Studies, 129
CGA, 208
Charged particles, 137-142
Chemistry, 256
Children, 106, 132-135
CHM cube, 59, 60
Choreography, 257

Cicero, 42
Cinema, 38-39, 190
Claris CAD, 231
Classification systems, 43
Classroom, 161
Client-server models, 61
Clinical decision making, 105
Coates, Del, 211
Cognition
 assumptions about, 37-38
 images in, 93
 problems of, 43-45
Cognitive psychology, 257
Cognitive science, 79-81
Collaboration, 246
 academia and industry, 211-215
 art and, 176-177
 benefits of, 214-214
 computer graphics education, 189-198
 vs. individualism, 189-191
 networks and, 180-185
 problems with, 182-184
 skill in, 191
 students and professionals, 178
 teleconferencing and, 175-177
Collaboration In Computer Graphics, 184-185
Collaborative performance, 185-186
Color, 109, 110, 168-170, 208
Color plots, 110
COLOS (Conceptual Learning of Science), 143-144
Combinatorics, 97
COMMETT program, 143-144
Common training platform, 55
Communication, 131
 intragroup skills in, 184, 185
 multimedia and, 29-30
 network and user, 50
 protocols for, 205
 skill in, 193
 teacher and learner, 49, 53-54
 technology for, 51
 user to user, 59
 *see also* Telecommunication
Compact Disc Interactive (CD-I), 25

# Index

Complexity, 138
Complex systems
  analysis of, 95, 97
  understanding of, 79, 82
Compound document grammar, 16
Compression, image, 261-262
Computer-aided design (CAD), 70, 135, 185, 227
Computer-aided industrial design (CAID), 211-215
Computer-aided instruction (CAI), 11, 65, 79, 250-254
Computer-aided lecturing, 254
Computer-based training, 54-56
Computer-generated sculpture, 184-185
Computer graphics
  art and, 112
  collaboration in, 189-198
  demonstrations of, 142
  education and, 103-113, 211-215, 217-225, 255-262
  education in, 65-77, 112-113, 189-198, 217-238
  future trends, 259-262
  hypermedia and, 65-77
  information provision by, 110-111
  insight from, 107-109
  interactive, 173-187
  interdisciplinary, 68, 70, 257
  solid modeling in, 227-238
  teaching and, 255-258
  telecommunications and, 180, 186
  video phones and, 175-176
  see also Visualization
Computer Graphics and Education '91, 1-5, 243-262
Computer industry, 25-26
Computerized Overhead Projector, 162
Computer languages, 145-156; see also specific languages
Computers, 209-210; see also specific products and systems
Computer Science, 228-230
Computer simulations, 93-94
Computers in Teaching intitiative, 253

Computer-Supported Cooperative Work (CSCW), 5, 55, 262
Computer technology, limits of, 107-109
Conference computing, 51, 53
Connections, 118
Constructive Solid Geometry methods, 70
Consumer electronics, 25-26
Content search, 16
Cooperation, *see* Collaboration
Cooperative client-server model, 61
Cooperative document access, 58-59
Cooperative hypermedia systems (CHM), 57-62
Cooperative learning, 49-62
Copyrights, 40
Costs
  computer systems, 201, 245-246
  hardware, 21, 96, 111
  software, 4, 96, 245, 257
  telecommunications, 187
Coulomb forces, 141-142
Course materials, 54
Courseware, 3-4, 161
  delivery of, 204-205
  LANs and, 206-207
  lifetime of, 208-209
  mathematics, 115-126
  portability of, 202-210
  prototypes of, 202-204
  standards for, 204-205
  *see also* specific programs and systems
Crayola, 22
Creative software, 132-136
Creative thinking, 179-180
Cross-reference links, 71, 72
Crystallographic groups, 122
Customizing software, 251

Database
  electronic books as, 10, 17, 23
  large, 109
  multimedia, 34
  networks and, 109
  remote access to, 207
  SIMPLE's, 84-85

Datasets, 32-33
Da Vinci, Leonardo, 98-99
DDE (Dynamic Data Exchange), 17
Decision making, 105
Delivery systems, 13
Demonstrations, 142
  algorithm simulator, 223-225
  courseware, 5, 202-204
Depth display, 221, 222
Design
  application interface, 123-124
  artistic, 112
  building, 227-228
  computer graphics and, 211-215
  course materials, 54
  educational software, 134-135
  education in, 211-215
  engineering, 184-185
  hypertext, 18
  language for, 230, 231, 236-238
  lecture, 162-168
  networks and, 184-185
  transparency, 162-166
Desktop metaphor, 43
Desktop supercomputers, 21
Dictation, 13-14
Digital video, 27, 34, 262
Directed graphs 14, 69
Displays
  abstract, 155-156
  algorithms for high resolution, 218
  ergonomics of, 11-12
  hypermedia and, 66, 67-68
  need for color, 109
  projected, 161-166
  reduction graph, 148-153
  screens vs. paper, 11-12
Distributed services, 61
DNA, 246
DNA of mathematics, 121
Docuchaos, 17
Documentaries, 244
Documents
  compound, 16
  CHM and, 57-62
  cooperative access to, 50, 58-59
  editing of, 58
  multimedia, 27-34, 49-51
Document-type definition (DTD), 16
Docuverse, 17
DQDB, 51
Drawing software, 166-168, 182-184, 231
Dream Machines/Computer Lib, 14
DTD (document-type definition), 16
DVI, 27
Dynabooks, 12, 23-24
Dynamical systems, 122
Dynamic insights, 81
DynaText, 15-16
Dyslexia, 98

Ease-of-use, 123
Ecology, 179
Economics, 190-191
Economy, information, 95
Edges, 219, 221-224
Editing, 58, 166-168
Education
  collaboration in, 189-198
  COLOS and, 143-144
  computer graphics, 65-77, 112-113, 189-198, 217-238
  computer graphics in, 103-113, 211-215, 217-225, 255-262
  individualism in, 190
  interactive illustrations and, 19-20, 22
  mathematics, 91-100, 115-126
  multimedia in, 25-26, 30-34
  research and, 261
  science, 91-100
  software companies and, 202, 251
  telecommunications and, 177-187
  trends in, 49, 51-55, 62, 91-100, 97-100
  university, 103-113
  verbal and numerical approaches, 93
  visualization in, 91-100
Educational computing, 103-104
Educational materials
  lifetime of, 208-209
  portability of, 201-210
Educational software, 132-136, 177, 250-254

cost of, 4, 257
funding of, 251-253
recommendations for, 252-253
*see also* Courseware
Educational video, 110-113
Einstein, Albert, 94-95
Eland, Joanne, 105
Electricity, 85-87, 137-142, 144
Electronic books, 9-24
The Electronic Cafe, 176, 180
Electronic Classroom, 177
Electronic courses, 54
Electronic libraries, 13
Electronic mail, 176-177, 206-207
Electronic publishing, 10
Electronic teacher, *see* Tutoring systems
Encyclopedia, on line, 207
Encyclopedia of Mathematics, 121
Engelbart, Doug, 13
Engineering, 98, 184-185, 228-230
Environmental issues, 190-191
Environmental Protection Agency (EPA), 195-196
EOP (Editor-Organizer-Presenter), 162-172
Equipotential lines, 141
Erasers, 133
Ergonomics, 11-12
Erlanger Program, 126
Error checking, 261
Euclidean crystallographic groups, 122
European community, 143-144, 253
Evaluation, 145-148
Evaluation history, 147-148, 154-155
Expansion links, 71, 73
Experience, 93, 99
Experimental Visual Techniques, 192-197
Experimenting, 131, 134
    demonstrations of, 142
    mathematical, 119
Expert systems, 248
Explanations, 248-249
Explorability, 29
Exploratory learning, 51, 117
Expressions, numerical, 145-146, 153-155
Externalisation, 247-249
Extrusion, 235, 236, 237

Eye pathology, 106

Faraday, Michael, 94, 95
Feedback, 135
Fiber optic cables, 175
Film-making, see Cinema
Filters, 147-156
Flexibililty, courseware, 203, 207
Fluent Technologies, 261-262
Fluid mixing, 107
Folberg, Robert, 106
Formats, mathematical object, 122
Four Freedoms, 115, 118-120, 124
Fourier analysis, 118, 142
Fractal geometry, 101
Freedoms, 115, 117-120, 124
Freehand, 192
Functional programming, 145
Functions, 145-156
Funding, 251-253

Gagn's objectives, 164
Galileo, 189
Galloway, Kit, 176, 180
Garbage storyboard, 195-197
Gardner, Howard, 100
Generalists, 98-99
Genius, 94-95
Genome Project, 246
Geographic data, 109
Geography, 179
Geometry, 92, 97, 260
Geometry Supercomputer Project, 125-126
GKS, 207, 261
Global Laboratory Project, 179
Godel's Incompleteness Theorem, 117
Go-to links, 71, 72
Gouraud shading, 75-76
Grammar, compound document, 16
Graphical editing, 231-234
Graphical User Interfaces (GUIs), 123
    solid modeling and, 230-238
    standards for, 201
Graphic computers, 91, 96, 206-208
Graphics
    automatic generation of, 209

EOP's, 167-168
  hypermedia and design of, 11-12, 18
  software for, 259-261
Graph structures, 57-58
Greek texts, 204
Grey-scale rendering, 208
Group learning, 54-56
Group projects, 70
Guide, 67, 71, 253

Hand-eye coordination, 130
Hard copy, 110-113
Hardware
  animation and, 21
  costs of, 21, 96, 111, 245
  future, 259-261
  hypermedia and, 66
  limitations of, 11-12
  multimedia and, 27
  needed, 27-28, 66
  standards for, 204-205
Hausman, Julie, 106
HDTV, 262
Hearing impairment, 106
Hemispherism, 94
Hewlett Packard, 111, 143
Hickman, Craig, 132-135
Hidden lines, 220-221
Hidden surfaces, 217-225
High frequency phenomena, 142
Holton, Gerald, 95
Human-computer interaction, 12
Human-computer interface, *see* Interface
Human Genome Project, 246
Humanities, 204
Huntley, John, 107
Hyperbooks, 202, 209
Hypercard, 11, 25, 26, 67, 253
  nursing education and, 105
HyperGraph, 65-77
Hyperlinks, 57
Hypermedia, 11-18
  assumptions about, 37-38
  authoring systems for, 67-68
  computer graphics education and, 65-77, 225

higher education and, 104
  mnemotechnics and, 42-43
  navigation of, 14, 17-18, 57-58, 65
  problems for, 38-42, 43-45
  simulation engine and, 45-46
Hypermedia assistants, 166
Hypersimulator, 79-87
Hyperstructure, 57-58
HyperTalk, 11
Hypertext, 11, 13-18
  current systems for, 67
  defined, 14
  design guidelines for, 18
  hypermedia and, 39
  intuition support by, 121-122, 124
  mathematics education and, 121-122, 124
  navigation in, 44, 121
  node views for, 69
  rhetoric of, 18
  SIMPLE and, 84-85
  *see also* specific systems
HyTime, 262

IBM
  digital video and, 27
  hypermedia and, 66-68
  multimedia and, 26
  PCs and compatibles, 66-68, 71, 204
  Prodigy system, 10-11
  workstations, 111
Icons, 180
Idealised forms, 138
Ideas, visualizing, 256
Identity filter, 150
Illumination model, 73
Illustrations
  biological, 189
  interactive, 19-24
Illustrator, 192
Images
  collections of, 41
  compression of, 261-262
  data for, 12
  processing of, 133, 180
  quality of, 109
Image space algorithms, 218-219

# Index

Imaginary numbers, 118
Imagination, 132-136
Imaging techniques, 193
Imitation, 93
Individualism, 189-191
Individual learning, 53
Industrial design, *see* Design
Industry, 191, 211-215
Informatica Grafica, 228-230
Informatics, 51, 209-210
Information
  access to, 23
  display of, 110
  multiplexing of, 38-39
  provision of, 110-111
  qualitative, 82
  serial presentation of, 44
  sharing, 43
  spatial, 81
  volume of, 40-43
Information economy, 95
Innovation, *see* Technology, change in
Input, 261
Insight, 81, 107-109
Institute of Architecture, Building and Town Planning, 227-228
Integration operations, 234-236
Intel, 27
Intellectual property, 23, 40, 214-215
Intelligence, 95
Intelligent links, 57
Intelligent Tutoring Systems (ITS), 70
Intelligent search, 18-19
Interaction, 123
  art and, 185-186
  computer graphics and, 217
  electronic assistance and, 245, 247-249
  illustrations and, 19-24
  representation and, 140
  with simulations, 108-109
  teaching and, 49, 53-54, 223-224
  telecommunications and, 173-187
  visualization and, 247-249
Interdisciplinary computer graphics, 257
Interdisciplinary education, 192

Interface
  abstract, 146-147, 155-156
  design of, 123-124
  human-computer, 43, 203
  network, 55-56
  standards for, 205
  virtual campus, 55-56
Intermedia system, 14
Internalizing, 130
International Advisory Group (INAG), 76
International cooperation, 253
International computer graphics usage, 255
Interoperability, 68
Intuition
  mathematical, 115-119
  software support of, 120-124
Invisible surfaces, *see* Hidden surfaces
ISDN (Integrated Services on a Digital Network), 51
ISO, 261-262

Joan Truckenbrod's Studio, 176
Jobs, 212-213; *see also* Work
Jobs, Steve, 205
Joined editing, 50
JPEG (Joint Photographics Expert Group), 28, 261-262

Kay, Alan, 12
Kearney, Joe, 110
Kernels, virtual campus, 55-56
KID PIX, 132-136
Kinesthetic thinking, 130
Knowbots, 19
Knowledge base, 12-13
Knowledge creation, 107

Landon, Brooks, 104
Languages
  human, 12, 258
  mathematical structure, 122
Lazy evaluation, 154-155
Lazy functional languages, 145-156
Learning, 23, 51-55
  cooperative, 49-62

current technology and, 250-254
difficulties with, 94-96
experience and, 93
hypermedia and, 49-62
hyper-simulator and, 79-87
imitation and, 93
mental maps and, 132
student-controlled, 247-249
visual, 3, 243-249
Lebanese University, 224
Lectures, 53-54, 65
computer asisted, 161-172
computer graphics and, 256
design of, 162-168
vs. experience, 93
hardware, 96
telecommunications and, 178
Le Prof, 254
Liberal arts, 191
Libraries, 9-10
electronic, 13
information location in, 41
Licencing, software, 252
Light, 73-76
Linking protocols, 17
Links, 57
freedom to make, 118
hypermedia, 14, 50
hypertext, 67, 71-73, 75
Lip reading, 106
Literacy, visual, 244-249
Local Area Networks (LANS), 177-178, 180-185, 206-207
Logic, mathematical, 115-119
Logical thinking, 244
London Transport Museum, 41
Lost in hyperspace, 17-18, 164

MacDraw, 182, 231
Macintosh, 25, 182, 201
algorithm simulator and, 224
costs of, 111
hypermedia and, 15, 17, 67-68, 71, 104
Minerva and, 230-235
REL and, 192
teleconferencing and, 175

Management, 55
Manipulative skill, 130
Mathematica, 31-32
Mathematics
art and, 194
education in, 91-100, 115-126, 142
experimental, 119
formalism in, 137
logic and intuition in, 115-119
MediaView and, 31-32
programming and, 145-156
visualization and, 91-100, 137, 194-195
Mathematics!, 257
Mathenesse, 233f, 234f
Math set, 134
Maturational lags, 95-96
Maxwell, James Clerk, 94, 95
MediaView, 18, 26, 29-34, 253
Medical simulations, 256-257
Medieval clerk, 98-100
Memex, 13-14, 104
Memory, 42-43, 130-132
Mental maps, 132
Mental models, 81, 99
Mental visualization, 130
Metamorphic imagery, 195
Metaphor limitations, 11-12
Method of loci, 42
Microfilm, 14
Microworld, 21
MIDI interface, 45
Minerva, 227, 229-235
Minimalist Instruction, 66
Miranda, 146-153
Mnemotechnics, 42-43
Model-driven animation, 20, 22
Modeling, 259, 260
solids, 260
three-dimensional, 110
Models, explanatory, 248-249
Modem, 176-177
Modifiability, courseware, 203, 207
Morphologically-based searching, 15
Morse, Samuel, 129
Motif, 143, 201
Motivation, 257

# Index

MPEG (Motion Picture Expert Group), 28, 261-262
MS/DOS, 201
Multi-dimensional representation, 139
Multi-lingual support, 4
Multimedia, 10-12
  CHM and, 58
  definition, 25
  education and, 25-26, 30-34, 180
  future abilities of, 260-262
  hypermedia and, 39
  lecturing and, 165
  limitations of, 11-12
  network for, 28, 173
  NeXT and, 205
  publishing, 253
  requirements for, 26-29
  technology for, 49-51
  telecommunications and, 180, 185-186
  term papers, 107
  university education and, 103-113
Multiple views, 82, 84
Multiplexing, 38-39
Muscle memory, 130
Music, 38, 39

Narrative style, 44
The National Geographic Kids Network, 179
National Science Foundation, 179
Natural language, 70, 248-249
Nature, 189
Navigation
  DynaText, 16
  evaluation history, 154-155
  HyperGraph, 69
  hypermedia, 14, 17-18, 57-58, 65
  hypertext, 44, 121
  lecture, 164
  paper books, 9
  tools for, 57-58
NCSA, 192-193
Negroponte, Nicholas, 25
Nelson, Ted, 13-14, 16-17
Network pal, 185

Networks
  digital video and, 27, 28, 34
  human interaction through, 50
  Knowbots and, 19
  local area, 177-178, 180-185
  long distance, 180-185
  multimedia, 173
  speed of, 108, 109
  telecommunications and, 51, 173
Neurological capacities, 93, 94, 96
NewSpeak, 19
Newton, 189
Newtonian physics, 21
NeXT, 18, 28-31, 205, 253
Non-termination, 153
Notes, visual, 247-249
NTSC, 262
Numerical Laboratory, 192
Numerical methods, 142, 145-146, 153-155
Nursing education, 105

Objective-C, 143
Objectives, 163-165
Object-oriented environments, 259-261
Object-Oriented Graphics Library (OOGL), 125-126
Object-oriented models, 84
Object-oriented programming, 122-124, 143-144, 207
Object space algorithms, 218
Office of Visual Materials, 106
Open Look, 201
Open systems, 15, 17
Operating systems, 205; *see also* specific systems
Ophthalmology, 106
Optical discs, 28
OS/2, 68, 201
Otolaryngology, 106
Oud's Mathenesse, 233f
Overhead transparencies, 161-166

Pain research, 105
Painting software, 133, 182-184

PAL, 262
Paper displays, 11-12
Paperless society, 10
Parallelism, 38
Parallel views, 233-234
Parametric surfaces, 232
Parametrization, 119
Particles, representation of, 139
Pasteur, Louis, 129
Pathology of the Eye, 106
Patient simulations, 105
Patterns, 130, 131, 134, 151, 153
Pearson, Karl, 91-92
Pedagogical points, 162-168
Pencil-and-paper evaluation, 146
Perception, 37, 38-40
Performance art, 185-186
Personalized environments, 82
Personal tutors, 96
Perspective theory, 234
Pestalozzi's principles, 94
PHIGS, 207, 261
PHIGS-like graphics, 231, 236
Philips Electronics, 25
Philip, Dick, 18
Phong method, 73-75
Photoshop, 192, 193
Physical modeling, 130
Physical sciences, 97
Physics, 21, 142, 259, 260
PLASM, 230, 231, 236-238
PLATO, 202, 250
Playacting, 130
Playing, 131, 134
Poincar, Henri, 116-117
Point charges, 137-142
Polygons, 218
Polyhedra, 230-238
Polyhedral approximation, 232
Polyhedral calculus, 236-238
Polyhedral networks, 220
Poprietary information, 214-215
Pop-up links, 71, 73
Portability, 17, 201-210
Postmodernism, 197-198
PostScript, 124

Power systems, electrical, 85-87
Presentation commands, 168-171
Presentation graphics, 110
Presentation methods, 244
Problem solving, 217
Prodigy system, 10-11
Product-design cycle, 212
Professional societies, 257
PROF mode, 171
Programming
  automatic, 209
  courseware, 203, 207
  efficiency of, 152-153
  graphics, 217-225
  intuition suport and, 122-124
  mathematics and, 122-124
  object-oriented, 122-124, 143-144, 207
  PLASM and, 236-238
  standards for languages, 205
Projections, 232-234
Projectors, 161-166
Property, intellectual, 23, 40
Prospero, 145-156
Prototypes, 212
  algorithm simulator, 224
  courseware, 202-204
Pseudo-coloring, 110
Publishing, multimedia, 253

Qualitative information, 82
Qualitative models, 137
Qualitative understanding, 80

Rabinowitz, Sherrie, 176, 180
RACE, 51
Real-time simulation, 46
Recursion, 153
Recycling, 196
Reddy, Raj, 12, 21
Reduced Instruction Set Computers (RISC), 205
Reduction graphs, 147-156
Remembering, 42-43
Renaissance, 189
The Renaissance Lab (REL), 192-197
Renaissance Man, 98-100

Renaissance Team, 198
Rendering, polyhedra, 235-236
Replacement links, 71, 73
Representation, 119, 138-142
Research
  computer graphics in, 103, 107-109
  courseware, 202-203
  education and, 261
  environments for, 261
  graphics methods and, 227
  individual vs. collaborative, 189-191
  mathematics, 116
Resolution, 208
Rhetoric, hypertext, 18
RISC, 205
RMG object-oriented programming environment, 143-144
Roaming, 118
The Robot, 175, 180
Robotics class, 110
Romboy Homotopy, 194
Root-Bernstein, Robert, 129-131
Rotation, 220-221
Roy, Chris, 106

Samuel, Arthur, 10
Satellite transmissions, 175, 178
Scale, assumptions about, 37
Scan line algorithm, 218-225
Scenario commands, 166-171
The School of the Art Institute in Chicago, 174, 179-180
Schools, 211-215; *see also* Universities; and specific institutions
Science
  art and, 257
  computers in, 203-204
  education in, 91-100
Scientific methodology, 189
Scientific visualization, *see* Visualization
Scientists
  artists and, 189
  visual thinking by, 129-131
Screw-extrusion, 235, 236f
Sculptors, 130
Sculpture, computer generated, 182-185

Search facilities, 15, 16, 18-19
Sears, 10-11
Second Look Computing, 104-107
Self-learning, 52, 254
Sminaire Bourbaki, 101
SGML (Standard Generalized Markup Language), 15-16
Shading, 224
Shading models, 75-76
Silicon Graphics, 111, 192
SILK (Speech, Image, Language, and Knowledge), 12, 24
SIMPLE, 81-87
Simple n/X, 230, 231, 235-236
Simplifying, understanding by, 80
Simulations, 93-94, 99, 256-257
  algorithm, 217-225
  real-time, 245
  engine for, 45-46
  interactions with, 108-109
  patient, 105
Site licences, 252
Situated understanding, 80
Skills
  collaborative, 191
  communication, 193
  manipulative, 130
  technological displacement of, 99-100
  workforce, 49
Slides, 110
Smalltalk, 156
Smithsonian Museum, 41
Software
  animation and, 22
  companies, 4, 202, 251
  costs of, 4, 96, 245, 257
  creative, 132-136
  customization of, 251
  design of, 134-135
  development of, 123
  educational, 4, 132-136, 177, 250-254, 257
  future, 259-261
  hypermedia and, 66
  intuition support with, 120-124
  lecturing, 162-172

licencing of, 252
limitations of, 26, 120
multimedia and, 26, 28-29
needed, 28-29, 66
telecommunications, 177
*see also* Courseware; Programming; and specific products
Solid modeling, 227-238, 260
Solid waste, 195-196
Sony Corporation, 175
Sorting algorithms, 152
Sound
　CAD and, 135
　hypermedia and, 38-40
　KID PIX and, 133
　video and, 262
Spatial information, 81
Spatial understanding, 80
Specialists, 98-99
Speech, 12
Spinnaker Plus, 67
SPRINT, 174, 175
Standard Generalized Markup Language (SGML), 15-16
Standards
　courseware, 204-205
　GUI, 201
　hardware, 204-205
　mathematical object, 122
　modem, 177
　multimedia, 261-262
　problems of, 12-13, 251
　3D graphics, 124
Stanley Collection of African Art, 106
Statistics
　art using, 196
　visualization in, 91-93
Stereo goggles, 23
Storyboarding, 193-197
Structure editing, 58
Structure graphs, 57-58
Student-controlled learning, 247-249
STUDENT mode, 171
Student model, 70
Student placement, 215

Students
　multimedia term papers of, 107
　telecommunications and, 186-187
Style sheets, online reading, 16
Sun, 111
Supercomputers, 21, 190, 252
Surfaces, hidden, 217-225
Syllabus, 163, 167
Synchronization, user request, 59

Talent, visual, 94-96
TANGO, 20
Targa Tips, 193
Teacher, role of, 182-184, 186-187
Teaching
　computer graphics and, 255-258
　programming, 151-153
　recognition of, 250
　science, 137-144
　*see also* Lectures
Teaching machines, 44
Teaching program, *see* Tutoring systems
Teamwork, see Collaboration
Technical Education Research Center (TERC), 179
Technology
　change in, 4-5, 201, 202, 208-209, 245
　communication, 51, 173-178, 187
　computer, 107-109
　learning and, 3-4, 250-254
　limits of, 107-109, 251
　multimedia, 49-51
Telecommunications, 25-26, 173-187
　costs of, 187
　current technology in, 173-178, 187
　networks for, 51
Teleconferencing
　EOP and, 171
　students and teachers and, 186-187
　video, 173-176
Telematics, 51
Telephone companies, 25-26
Telespace, 185
Television, 190
　editors for, 42

# Index

hypermedia and, 38-40
news, 244
Term papers, 107
Tesla, Nikola, 99
Textbook, EOP as, 171-172
Text searching, 15
Thermodynamics, 92
Thesaurus Linguae Graecae, 204
Thinking
  creative, 179-180
  tools for, 129-131
  transformational, 131
  *see also* Understanding
Three-dimensions
  animation with, 193
  graphics with, 192, 217-225
  intuition support by, 124
  mathematics education and, 124
  modeling of, 110, 184
  standards for, 124
  viewing tools, 125
3G Machines, 12-13, 21, 258
3T Machines, 258
Tiles, 97
Timbuktu, 175, 181-184
Time dimension, 20-21
Timing tools, EOP's, 170-171
Tools
  classification of, 59, 61
  development, 257-258
  drawing, 166-168, 182-184
  editing, 166-168
  educational software and, 257-258
  EOP's, 166-168, 170-171
  hypermedia, 59
  image processing, 133
  interaction, 22
  painting, 182-184
  personalization of, 82
  sculpting, 182-184
  SIMPLE's, 84
  software, 181-184
  structure visualization, 57-58
  thinking, 82
  3D viewing, 125

timing, 170-171
virtual campus, 56
visualization, 59, 95
Topas, 184
Training, 51-55
  changes in, 99
  computer-based, 54-56
Transformational thinking, 131
Translation, human language, 12
Transparencies, 161-166
Tree structures, 57-58
Tutoring systems, 51-53
  costs of, 96
  intelligent, 70
  links in, 57
  personal, 96
Tye-Murray, Nancy, 106

Understanding
  hyper-simulator and, 79-87
  visual aids for, 244-249
  ways of, 79-81
United Arab Emirates University, 224
Universities
  Alias Research and, 215
  COLOS and, 143-144
  computer graphics in, 103-113, 256-258
  educational software and, 250-254
  LANs and, 206-207
  technology funding by, 252
  *see also* specific institutions
University of Geneva network, 206-207
UNIX, 193, 201, 205, 207
  authoring systems for, 68
  HyperGraph and, 71
  hypertext and, 68
  Mathemetica and, 31-32
  MediaView and, 31-32
  RMG and, 143
Unrolling, 165, 168-170
User, 19-24, 26

Venus & Milo, 194-195, 197
Verbal difficulties, 94-95
Verbal education, 93

Version control, 17
Video
  analog, 27
  demonstrations of, 142
  digital, 27, 262
  editors for, 42
  educational, 110-113
  multimedia and, 50
  teleconferencing, 173-176
Video conferencing, 34
Videodiscs, *see* CD-ROMs
Video phones, 175-176
Videotex, 10
Views, parallel, 233-234
Virtual campus, 55-56, 62
Virtual reality, 22-23, 185, 259
Visualization, 68, 194-195, 256
  collaboration in, 190
  education and, 91-100, 119
  hypermedia structure, 58, 59
  interactive, 247-249
  mental, 130
  Physics and, 137-144
  scan line simulator, 220-225
  tools for, 95
Visualization Program, 192
Visual literacy, 3, 244-249
Visual maps, 132
Visual notes, 247-249
Visual study, 189
Visual talents, 94-96
Visual thinkers, 91-96

Visual thinking, 124, 129-131
Voice dictation, 13-14
Voice recognition, 13-14
Voltage, 85-86
VP system, 155-156

Wavefront Technologies, 193
Webs, 15
Weeg Computing Center, 105
Weiner, Norbert, 99
Windowing systems, 123
  abstract, 155
  DDE in, 17
  hypertext using, 67-68, 253
  Microsoft Windows, 17, 68, 71
  Motif, 143, 201
  solid modeling and, 230-235
  *see also* Macintosh; NeXT
Winged-triangle, 231, 235
Wire frame pictures, 223-224
Word processor metaphor, 29-30
Work, 49, 93-94, 99, 212-213
Workspace Manager, 205
Workstations, 111
  animation and, 21
  design and, 212
Wright, Frank Lloyd, 107

Xdico, 207
X3J3 committee, 205
X Windows, 68, 143, 207

Printing: Mercedesdruck, Berlin
Binding: Buchbinderei Lüderitz & Bauer, Berlin